791476
7/77

£17.00

BIRMINGHAM UNIVERSITY LIBRARY

~~ST BE RETURNED
~LLED FOR
~WER,

D1336743

SKEW DISTRIBUTIONS AND THE SIZES
OF BUSINESS FIRMS

STUDIES
IN MATHEMATICAL AND
MANAGERIAL ECONOMICS

editor

HENRI THEIL

VOLUME 24

NORTH-HOLLAND PUBLISHING COMPANY
AMSTERDAM · NEW YORK · OXFORD

SKEW DISTRIBUTIONS AND THE SIZES OF BUSINESS FIRMS

YUJI IJIRI and HERBERT A. SIMON

Carnegie-Mellon University

Collaborators

CHARLES P. BONINI   THEODORE A. VAN WORMER

1977

NORTH-HOLLAND PUBLISHING COMPANY
AMSTERDAM · NEW YORK · OXFORD

© 1977 North-Holland Publishing Company

All rights reserved. No part of this publication may be reproduced, stored in a retrieval system, or transmitted, in any form or by any means, electronic, mechanical, photocopying, recording or otherwise, without the prior permission of the copyright owner.

Library of Congress Catalog Card Number: 76-20682
ISBN North-Holland: 0 7204 0518 1

PUBLISHERS:

NORTH-HOLLAND PUBLISHING COMPANY
AMSTERDAM · NEW YORK · OXFORD

DISTRIBUTORS FOR THE U.S.A. AND CANADA:

ELSEVIER NORTH-HOLLAND, INC.
52 VANDERBILT AVENUE
NEW YORK, N.Y. 10017

Library of Congress Cataloging in Publication Data

Ijiri, Yuji.
Skew distributions and the sizes of business firms.

(Studies in mathematical and managerial economics;
v. 24)
Bibliography: p. 219
Includes index.
1. Industries, Size of – Mathematical models.
2. Distribution (Probability theory) I. Simon,
Herbert Alexander, 1916–   joint author. II. Title.
III. Series.
HD69.S5I45   1976   338.6'4   76-20682
ISBN 0-7204-0518-1

PRINTED IN THE NETHERLANDS

INTRODUCTION TO THE SERIES

This is a series of books concerned with the quantitative approach to problems in the social and administrative sciences. The studies are in particular in the overlapping areas of mathematical economics, econometrics, operational research, and management science. Also, the mathematical and statistical techniques which belong to the apparatus of modern social and administrative sciences have their place in this series. A well-balanced mixture of pure theory and practical applications is envisaged, which ought to be useful for universities and for research workers in business and government.

The Editor hopes that the volumes of this series, all of which relate to such a young and vigorous field of research activity, will contribute to the exchange of scientific information at a truly international level.

THE EDITOR

Preface

We are grateful to the following journals and publishers for permission to reprint, with modifications, the articles from which the chapters of this book are drawn:

American Economic Review

"The Size Distribution of Business Firms," 48: 607–617 (1958). (Chapter 7)
"Business Firm Growth and Size," 54: 77–89 (1964). (Chapter 8)

Behavioral Science

"Some Monte Carlo Estimates of the Yule Distribution," 8: 203–210 (1963). (Chapter 5)

Biometrika

"On a Class of Skew Distribution Functions," 52: 425–440 (1955). (Chapter 1)

Econometrica

"A Model of Business Firm Growth," 35: 348–355 (1967). (Chapter 9)

Information and Control

"Some Further Notes on a Class of Skew Distribution Functions," 3: 80–88 (1960). (Chapter 2)

Journal of Political Economy

"Effects of Mergers and Acquisitions on Business Firm Concentration," 79: 314–322 (1971). (Chapter 10)
"Interpretations of Departures from the Pareto Curve Firm-Size Distributions," 82: 315–331 (1974). (Chapter 11)

North-Holland Publishing Company

"On Judging the Plausibility of Theories," in B. van Rootselaar and J. F. Staal (eds.), *Logic, Methodology and Philosophy of Sciences*, Vol. III (1968). (Chapter 6)

Proceedings of the National Academy of Sciences

"Some Distributions Associated with Bose–Einstein Statistics," 72: 1654–1657 (1975). (Chapter 4)

The original articles have been left largely intact, with alterations limited to changes in symbols and style, to make them consistent throughout the book. At the expense of a little duplication in the opening portions of some chapters, which results from keeping them in their original form, the individual chapters can be read more or less independently of each other.

Each chapter begins with a brief introductory note that indicates its relations with those that precede and follow it, while a new general introduction gives an overview of the unifying theme that threads through the entire volume.

We are indebted to C. P. Bonini and T. A. Van Wormer, coauthors of chapters 7 and 5, respectively, as well as to numerous other colleagues whose help is noted in footnotes to individual chapters. Our work has been supported in part by Grant GS-33355 from the National Science Foundation, which we acknowledge with gratitude.

Y. Ijiri
H. A. Simon

Contents

Introduction

Nature, as it presents itself to the physical scientist, is full of clearly defined patterns. The planets move in ellipses about the sun. The distance traveled by a falling body varies with the square of the time. The current in a wire varies inversely with the resistance and directly with the voltage. The volume of a gas is proportionate to its absolute temperature and inversely proportionate to its pressure. These are only the simplest regularities the physicist observes. The nineteenth century brought forth a whole array of partial differential equations to describe in exquisite detail the motions of waves: light, sound, electromagnetism. And twentieth century physics has discovered even more elaborate and arcane patterns in the architecture of the atom and in the groups and symmetries derivable from quantum mechanics.

The patterns that have been discovered in social phenomena are much less neat. To be sure, economics has evolved a highly sophisticated body of mathematical laws, but for the most part, these laws bear a rather distant relation to empirical phenomena, and usually imply only qualitative relations among observables: the quantity of a commodity purchased decreases as the price rises; if a tax is placed on a commodity, its price will rise by a fraction of that tax. One of the most important contributions to modern mathematical economics, Samuelson's celebrated essay on "Comparative Statics and Dynamics," devotes itself to showing how the direction of movement of economic variables from one static equilibrium to another can be deduced from assumptions about the dynamic stability of the economic system. Thus, we know a great deal about the direction of movement of one variable with the movement of another, a little about the magnitudes of such movements, and almost nothing about the functional forms of the underlying relations.

Hence, on those occasions when a social phenomenon appears to exhibit some of the same simplicity and regularity of pattern as is seen so commonly in physics, it is bound to excite interest and attention. Those occasions have been rare. Some economists have concluded from the evidence that the price level does, indeed, vary proportionately with the quantity of money; but the pattern is sufficiently difficult to extract from the noisy data that its existence can be, and is, disputed. The marginal propensity to consume has been thought by some economists (mostly the same ones who accept the quantity theory of money) to be an invariant, independent of income. The proposition is tenable, but it hardly leaps forward from the data to strike the observer. Rather, it has had to be teased out by painstaking statistic tests that conclude that "the hypothesis that the marginal propensity to consume is constant cannot be rejected."

This book is concerned with a regularity in social phenomena that is both striking and observable in a number of quite diverse situations. It is a regularity in the size distributions of various kinds of social aggregates: of personal wealth, of incomes, of business firms, of cities, of publication frequencies, and of word frequencies. All of these distributions fit quite closely a Pareto distribution. If, for example, the cities of the United States are ranked according to size from large to small, then the logarithm of size will be seen to be negatively and linearly related to the logarithm of rank. If the families in the United States are ranked according to wealth, from wealthiest to poorest, then the logarithm of wealth will vary linearly with the logarithm of rank. And so for the other examples: the distribution of incomes, of business firms according to almost any measure of size, the distribution of scientists according to the number of papers they have published, or the distribution of different words in James Joyce's *Ulysses* according to the number of times they occurred.

Our objective in bringing together the material in this book is twofold: first, to present some general theory to account for this persistent regularity in rank-size distributions; second, to apply the theory specifically to the data on firm size distributions, and to draw out the economic implications, including implications for policy, from the application. The first six chapters are devoted to the general analysis of skew distribution functions of the observed kind; the last five chapters are devoted to the analysis of the size distributions of business firms.

1. Skew distribution functions

It is argued in the first chapter of this book that if the very same regularity appears among diverse phenomena having no obvious common mechanism, then chance operating through the laws of probability becomes a plausible candidate for explaining that regularity. Hence, the theoretical models we shall examine are stochastic models yielding the observed size distributions as their steady-state equilibria.

With each stochastic model are associated particular assumptions about the probabilities associated with its elementary processes. By making different assumptions about those probabilities, one can obtain any number of familiar or unfamiliar equilibrium distributions. Now the crucial task is to make the appropriate assumptions.

"Appropriateness" means two distinct things here. In the first place, assumptions must be chosen that will yield an equilibrium distribution resembling the observed one. That goes without saying. The second, and even more critical, criterion of "appropriateness" is that the underlying assumptions of the model should provide a plausible explanatory mechanism for the phenomena.

The cumulative distribution function in which the relation between the frequency and rank is linear on a logarithmic scale is usually called the Pareto distribution. Another distribution that approximates the Pareto in its upper tail is the Yule distribution. We shall be mainly concerned with these two distributions and with others that resemble them. Two of the others we shall consider from time to time are the log normal distribution and Fisher's log series distribution.

The most prominent characteristics of this family of distributions is that they are highly skewed in their upper tails. The kind of stochastic assumption that leads to such skewed distributions, the Gibrat assumption, postulates that the expected percentage rate of growth in size is independent of the size already attained.

The way in which we will therefore answer the first question of the appropriateness of the stochastic assumptions is to assume probabilities of growth that satisfy Gibrat's assumption, or something close to it. The second question of appropriateness will be answered by seeking interpretations of Gibrat's law of growth that make sense for the phenomena we are seeking to explain. In the case of business firm sizes, these interpretations will amount to some kind of assumption of constant returns to scale, for if the profitability of a business firm is independent of its current size, then we would expect average growth rates independent of size, in conformity with Gibrat's assumption.

A great deal has been written, at one time or another, as to whether a particular empirically observed distribution could, or could not, be approximated by the Yule distribution, the Pareto or the log normal. Such a question is difficult to answer, for several reasons. One reason, which is discussed at length in ch. 6, is that there does not really exist a body of statistical theory that tells one whether some data fit a particular extreme hypothesis.

The difficulty is that, with sufficiently large samples (and most of the samples we shall be dealing with are quite large), the chi-square test will almost always reject the hypothesis that the observed data agree with those calculated theoretically. The reason for this is that our theories are always only approximate theories that do not capture all the fine structure of the phenomena. Hence, with sufficiently large samples of sufficiently good data, the deviations of data from theory almost always reveal themselves. However, we cannot conclude from this that the theory should be rejected; the only valid conclusion to be drawn is that the theory is only a first approximation –

hardly surprising – and that the next step in the investigation is to look for an additional mechanism that could be incorporated in the theory so as to lead to a better second approximation. It would be foolish for us to give up the gas laws for ideal gases simply because most gases are not, in fact, ideal; or to give up the law of acceleration in a vacuum because most of the bodies we observe are falling through air.

Hence, we shall not be much concerned, in what follows, with significance tests, which are completely inappropriate for testing the fit of data to extreme models. Instead, we will be concerned with how much of the variance in the raw data is explained by the models, and with how sensitive the fit is to changes in assumptions. We will start out with the simplest possible models, in the interest of parsimony, and then introduce additional complications, either to see whether the fit of the data is much affected by detailed changes in the underlying assumptions, or to see whether second-order deviations of data from theory can be explained by auxiliary hypotheses. Above all, we shall try to keep in mind that the goal of all such work is to discover valid explanations for phenomena in the real world, and that statistical techniques, if they are of any use at all, are only of use to that end.

Finally, this book does not aspire to make contributions to the mathematics of stochastic processes; although a few modest advances will be found along the way (see especially chs. 3 and 4). We approach the mathematics as mathematical economists interested in empirical phenomena, and not as mathematicians. The main features of the data with which we are dealing can be understood in terms of steady-state expected values of the stochastic processes. Hence, our principal mathematical procedure will be to derive necessary conditions for a steady-state equilibrium, without treating in detail the diffusion processes that lead to the equilibrium. This is not only the easier path, but it is also a prudent one, for in all but the simplest cases closed solutions for the diffusion processes may not be available. As a safeguard against insufficient rigor leading to incorrect conclusions, we have done a considerable amount of simulation (see, for example, ch. 5) of the particular stochastic models we are concerned with. We obtain thereby a direct test of whether the systems will actually arrive at, and remain at, the steady state equilibria that are determined by the mathematical analysis. We think that in this way we are able to carry through an adequate analysis with a minimum of mathematical machinery and pyrotechnics.

2. Firm-size distributions

The chapters in the second part of this book undertake to explain the observed facts about the size distributions of business firms in terms of the stochastic models of Part I. This approach represents a complete departure from classical economic theory on this subject, which either ignored these distributions or sought to explain them in terms of the shape of the long-run cost curve.

The main outlines of the stochastic theory are given in ch. 7. The remaining chapters seek to explain the "fine structure" of the phenomena, by increasing the realism of the assumptions incorporated in the stochastic models and by trying to account for small departures of the observed distribution from the predictions of the theory in its simplest form. In this Introduction, we will expand on several of the themes that run through these five chapters.

During the period in which these chapters were first written (1958 to 1974), a number of substantial studies of firm size distributions appeared, among them the works of Steindl (1965), Singh and Whittington (1968) and Wederwang (1965). These books provide important empirical tests, to which we will refer in this introduction and the introductions to the individual chapters, of some of the assumptions of our models.

2.1. Classical and stochastic theories of firm size

The relation between size of firm and costs has always been of some interest to economic theory, although never a central topic. The interest derives from the fact that decreasing returns to scale (increasing marginal cost) is an essential condition to the existence of a competitive equilibrium.

The facts, however, do little to satisfy this requirement of theory. Walters (1963), who has carried out a comprehensive survey of empirical estimates of cost functions, examines some 34 such estimates. Half of these refer to railroads and utilities, and show constant or increasing returns to scale, as might be expected. Six of the remainder are estimates of long-run cost curves, and only one of these, even under the most liberal interpretation of the data, exhibits the U-shaped curve demanded by standard economic theory. Likewise, of the 12 short-run cost curves, only three show any signs of increasing marginal costs (Walters (1963, pp. 48–50)).

A person who was not so deeply attached to neoclassical theory as to assign an overwhelming prior probability to the hypothesis of decreasing returns to scale would have to conclude from the data that the evidence favors constant returns to scale and offers no support for U-shaped long-run cost curves. However, economics is not a discipline in which hypotheses that follow from classical assumptions, or that are necessary for classical conclusions, are quickly abandoned in the face of hostile evidence. Walters, for example, after reviewing the evidence and all the theoretical arguments that could be found to challenge it, concludes (page 51) that "the evidence in favour of constant marginal cost is not overwhelming. Certainly the revision of theory to include this phenomenon is not an urgent matter." A person with a more symmetrical distribution of his prior probabilities would reach a quite different conclusion: that an assumption of constant returns to scale is most consistent with the data.

Some further empirical support is lent to the latter conclusion by the general finding (see especially, Singh and Whittington

(1968, ch. 6) and Steindl (1965, pp. 226–228) for the evidence) that there is little or no correlation between size of firm and rate of profit. Of course this evidence, by itself, says nothing conclusive about the shape of the cost curve. It does show that firms over a large size range operate at comparable profit rates. If that were not so, the Gibrat assumption of equal expected growth rates would be quite implausible.

While the theory of competitive equilibrium is not compatible with constant costs, the theory provides no very convincing arguments (other than this one from necessity) to show why a business firm should experience decreasing returns to scale. A distinction must be made, of course, between the scale of the firm and the scale of its individual plants. It is not hard to think of technological reasons why there might be, at any moment in history, an optimum size of plant in a particular industry, average costs being higher for plants that were either larger or smaller than the optimum. If the technology of production did not impose a limit on size, such a limit might be fixed by the size of the markets that could be served efficiently from a single location (but notice that this latter consideration requires an elaboration of the theory beyond the usual assumptions of perfect competition).

Since there is no limit, however, on the number of plants a single firm can own and operate, limitations on the size of plant carry no implications for the optimum size of the firm. Hence, theorists have had to invent scarce factors associated with the firm (the surrogates for the "land" of Ricardian theory) whose supply could not be increased without sacrifice of quality, or could be increased only at increasing cost. A candidate frequently nominated for the scarce-factor role is "managerial talent." Another is "imperfect capital markets." Notice that these solutions always carry us into the theory of imperfect competition or of rents, but that is not a fatal objection. More serious is the fact that no one has produced empirical evidence that such scarcities actually exist or that, if they exist, they account for the observed limits on firm sizes.

A still more serious difficulty for the classical theory is that, even if we admit decreasing returns to scale, no predictions follow for the nature of the firm size distribution. The classical theory would admit a normal distribution, a rectangular one, or a single size for all the firms in an industry as readily as it would admit the skew distributions (whether they be Pareto, Yule or log normal) that are actually observed.

The classical theory describes the technology in terms of a family of short-run cost curves representing the costs associated with firms of different sizes. Once the size of the firm is determined (i.e. its short-run cost curve chosen), it will operate at the point on that curve where marginal cost equals marginal revenue. Prior to determining its size, however, the opportunity set for the firm consists in the envelop of the family of short-run curves; this envelop is the long-run cost curve (fig. 1). The minimum of the envelop defines the optimal size of firm.

The venerable myth of the long-run cost curve is repeated solemnly in most of the standard economics textbooks as well as the books on intermediate price theory. The accompanying figure always depicts a properly U-shaped envelop, so that, in spite of contrary facts, the belief in decreasing returns to scale is not questioned.

The classical theory says nothing about whether the same cost curves are supposed to prevail for all of the firms in an industry, or whether, on the contrary, each firm has its own cost curve and its own optimum scale. If the former, then all firms in the

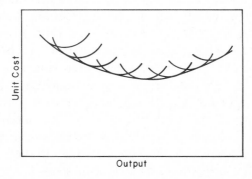

Fig. 1. The short-run and long-run cost curves, according to classical theory.

industry should be the same size. A prediction could hardly be more completely falsified by the facts than this one is. Virtually every industry that has been examined exhibits, instead of a clustering of its firms around a modal size, a highly skewed distribution of sizes with very large firms existing side by side with others of modest size.

If each firm, on the other hand, has its own peculiar optimum, then the theory says nothing about what the resulting distribution of these optima for the industry should be. It certainly provides no explanation for the repeated appearance of distributions resembling the Pareto. Thus, the theory either predicts the facts incorrectly or it makes no prediction at all.

These comments do not exhaust the list of difficulties from which the classical theory suffers. If it fails to explain why the firm sizes in an industry should satisfy the Pareto distribution, it equally fails to explain why the sizes of the large firms in an entire economy should fit that distribution – as the facts show they do.

Classical economic theory is sometimes defended on the ground that, even if its assumptions are not true, the economic world behaves "as if" they were (Friedman (1953, p. 40)). Even this argument fails to support the standard theory of the long-run cost curve, for there are no observable consequences of that theory for the economic world. Without positive reasons for accepting it, and with the failure to find direct empirical evidence for decreasing returns to scale, there seems to be little reason why the theory should be preserved in the corpus of economic doctrine.

Enough has been said, perhaps, about what is wrong with the classical theory. On the other side, one of the positive advantages of the stochastic theory to be developed here is that it does not have to struggle with the difficult concept of "industry." Even if we set aside the problem of drawing boundaries around a group of commodities in order to differentiate one industry from another, we observe that business firms in our society limit their activities less and less to such homogeneous collections of commodities. It is reassuring, therefore, that the stochastic models can be used both to explain size distributions in the groups of firms that we conventionally call "industries" (as in ch. 7), and the much more heterogeneous collection of the firms of an entire economy (with which chs. 9 through 11 are concerned). The model of ch. 9 perhaps shows most clearly how industries can be accommodated in the models or, alternatively, ignored.

2.2. *Proceeding by successive approximations*

In §1 of this Introduction, reference was made to two different criteria in terms of which the appropriateness of a theory to a body of data could be judged. On the one hand, the fit of the theory to the data must be considered. On the other hand, the plausibility of the explanatory assumptions on which the theory rests must be weighed. These methodological issues are explored at some length in ch. 6.

Accepting the methodological viewpoint of Part I, we have approached the analysis of the data on business firm sizes with the aim of reaching an understanding of these data by successive approximations. We start (e.g. in ch. 7) with a very simple stochastic theory, drawn from Part I, and fit it to some data. Then we notice two kinds of difficulties:

(1) When viewed closely, the assumptions of the theory contain unrealistic components. First, the simplest stochastic models that lead to the Pareto or Yule distributions do not admit temporal serial correlation of growth rates. (Empirical evidence for such serial correlation will be found, for example, in Singh and Whittington (1968, ch. 5).) Second, they ignore the phenomena of mergers and acquisitions. In the remaining chapters of Part II, we have tried to see how far we could go in removing these unrealistic restrictions from the theory. Chs. 8, 9 and part of 11 are concerned with serial correlation in growth; while ch. 10 and the remainder of 11 deal with mergers and acquisitions.

(2) When the fit between theory and data is examined, systematic differences can be seen between the theoretical and empirical curves. The most important of these is that the empirical data, when plotted on log–log paper, almost always exhibit a noticeable concavity toward the origin. The Pareto distribution is linear on a log–log scale. Chs. 8 and 11 suggest modifications of the basic theory that would produce non-linearities like those observed.

In general, it has turned out that the changes in the theory called for in order to make its assumptions more realistic also alter its predictions in directions that bring them into better agreement with the data. This, of course, is what we would hope from a strategy of successive approximations.

There are two other possible causes of concavity toward the origin that are not treated in chapters of Part II, but which are dealt with in Part I: the first of them in ch. 1, and at length in ch. 5; the other in ch. 2.

Translating into the language of business firm sizes, it is shown in chs. 1 and 5 that if the rate at which new firms enter is not constant, but decreases gradually over time; then, provided that the change in rate of entry is not too rapid, the data will continue to approximate the Pareto distribution, but with something of a concavity toward the origin. We have not actually examined time series data on rates of entry of business firms, but such an investigation might reveal the cause of part of the concavity of the observed data that is not explained by the other components of our model.

In ch. 2 it is shown that a concavity toward the origin could also arise if the economy were not growing but increases in the sizes of some firms were compensated by decreases in the sizes of others, both increases and decreases obeying the Gibrat assumption. Some further comments will be made about firm size distributions in a static economy at the end of this Introduction.

The stochastic models of firm size distributions can be tested in other ways than simply by comparing their steady-state equilibria with the observed distributions. By using time series data and the transition matrices describing the changes in size from one point in time to another, some of the basic postulates of the model can be tested directly and the parameters estimated. For example, in ch. 9, we make direct estimates from the empirical data of the serial correlation of growth rates. The analysis of mergers and acquisitions in the final two chapters is based on direct evidence of merger frequencies for firms of different sizes.

Although transition matrices were introduced in the pioneering 1956 study of Hart and Prais, they have not received nearly as much attention as the equilibrium distributions. Interesting data on the transition matrices for British firms will be found in ch. 5 and Appendix E of Singh and Whittington, as well as estimates of the amount of autocorrelation of growth over time.

2.3. Implications for public policy

In discussions of industrial concentration and anti-trust policy, it is commonly asserted or assumed that there is an inverse relation between the usual indices of concentration and the effectiveness of competition. The actual relation is far from obvious.

In the first place, frequently used measures of concentration, like the Gini index, or the fraction of total industry sales that are accounted for by some fixed number of largest firms, do not have any clear theoretical foundation. It would seem more defensible to measure concentration by a parameter of the stochastic process that is being used to explain the data – for example the slope of the Pareto curve. If this measure be used, then the degree of concentration in large-scale American industry has changed very little over a considerable period of years (See ch. 10). At least two claims can be made for the slope parameter of the distribution as a relevant measure of the effectiveness of competition.

First, a stable slope of the Pareto curve means that the large firms have been growing no more rapidly, on average, than their smaller competitors. This does not mean, of course, that *very* small firms are faring as well, for the total percentage share of market of all firms above some cut-off point may be increasing. Second, a stable slope of the Pareto curve, combined with linearity of the log–log plot indicates that the percentage of total growth accounted for by new firms entering the system (or growing beyond some minimum cut-off size) is relatively constant. As we shall see, in fact the plotted distribution is usually not quite linear, and the typical concavity toward the origin may provide some evidence of a declining rate of new entry (although other possible explanations for the concavity will be provided as well).

As is pointed out in ch. 7, one might entertain still other conceptions of what constitutes effective competition. For example, competition might be judged viable to the extent that new firms grow rapidly and replace older firms in the upper ranks of the distribution (see Singh and Whittington, ch. 5). There is no necessary connection between the slope parameter, which depends mainly on rate of entry, and the stability or instability of the rankings of individual firms over a term of years. The latter may be measured by the dispersion of columns of the transition matrix, that is, the extent to which the growth rates of individual terms deviate from the expected value. Such a measure of the differential mobility of firms in the ranking can be thought of as an estimate of "equality of opportunity" to become large, something rather different from the more usual static measures of equality or inequality of size.

If there are frequent and sizeable changes in rank, we may conclude that there is vigorous competition in the industry or economy. If the ranks of firms change infrequently, we may conclude that there is little competition. Although stability of rank does not necessarily imply lack of competition, it provides a plausible indication of it.

As a specific indicator we may use the relative rank, $q_i = r_i / r^*_i$, where r_i is the rank of the ith firm at the end of a period, and r^*_i is its rank at the beginning of the period. We would regard a ten percent change in rank as equally significant regardless of whether it is a change from 10th to 9th or from 100th to 90th in the population. If the population consists of the same n firms at the beginning and end of the period, then $\Sigma \log q_i = 0$. Even when this condition does not hold, the standard deviation of q, σ_q, may be a suitable indicator of the average amount of shifting in rank during the period.

We have carried out some sample calculations of σ_q using the *Fortune 500* data for successive years. Our calculation shows that for 1965–66, $\sigma_q = 0.0618$ and for 1971–72, $\sigma_q = 0.0591$. A similar calculation for the 100 largest industrial firms in Japan for 1969–70 yields $\sigma_q = 0.0587$, which is remarkably close to the U.S. figures.

Singh and Whittington used some slightly different measures to determine the extent of rank changes among British firms in three industries over the period 1948–1960. The first of these measures was the probability that a randomly selected firm of a given size would overtake a randomly selected firm of double that size. These probabilities ranged from 0.32 to 0.38 for the 12-year period. The same authors also computed rank correlations between the firm rankings at the beginning and end of the period.

In the present state of our knowledge, we have no theoretical basis for preferring one of these measures to another, nor do we have any standard for deciding what is "much" or "little" mobility in rank. This whole aspect of the growth of firms needs much additional study.

Finally, the magnitude of the autocorrelation of growth rates might be used as an (inverse) index of effectiveness of competition. This is so because the model to be developed in ch. 10 suggests that this serial correlation is a measure of the extent to which particular firms have temporary monopoly access to factors that allow them to increase their share of market. In classical economic theory, the fundamental measure of effectiveness of competition is how close the observed market equilibrium remains to the competitive equilibrium. The customary measures of concentration are simply surrogate indices, which have no theory-based relation to the phenomenon they purport to measure. In this respect, it must be admitted that the parameters of the stochastic models have no better claim than the conventional indices to measure distance from the competitive equilibrium. If the stochastic models do, indeed, give a reasonable account of the dynamics of firm-size distributions, then there is serious need for research that will tie those models back to the classical theory of competitive equilibrium. This book can make no claim to forge that connection beyond one rather speculative link.

The stochastic models of firm size and growth are perhaps best described as postulating something close to neutral equilibrium among firms. All firms have the same expected rate of growth, and differences in actual growth rates are the result of stochastic elements that advance or retard the growths of particular firms for a longer or shorter period of time (see especially ch. 9). The neutral equilibrium is consistent with the apparent absence of a relation between profit rates and firm sizes (Singh and Whittington, ch. 6). If expected profit rates tended to vary with attained rates, but also independently of size, then firms could acquire incremental capital in proportion to their current sizes. (For some evidence of a strong relation between profit rates and growth rates, see Singh and Whittington, ch. 7.) Thus, approximation of the observed distribution to the Pareto could be taken as evidence of a not-large departure from competitive equilibrium. But this is only a sketch of the kind of argument that would have to be developed to provide a genuine connection with the traditional theory.

2.4. Size distributions in a no-growth economy

The models discussed in this book have been developed on the assumption that the economy as a whole is growing. Today, the possibility of an economy without such growth is a serious topic of discussion, and we look forward to a future when the population will grow slowly or not at all. It is therefore of some interest to consider how the conclusions of our analysis would have to be modified to fit a nongrowing world.

The third section of ch. 2 provides a basis for such a modification. In that section a model is presented of a rather general birth and death process which, by appropriate choice of parameters, can be made to represent the steady state of an economy in which the net gains in sales (or some other measure of size) of some firms just balance the net losses of the others. As is shown in the introduction to ch. 2, when the parameters are fixed in this way, the system will have as its steady-state equilibrium Fisher's log series distribution rather than the Yule distribution. Let $f(i)$ be the proportion of firms of size i. The probability density function for Fisher's log series distribution is $f(i) = [-\log(1 - 1/\rho)]\rho^{-i}/i$, where $\rho = 1/(1 - \alpha)$ and α is the birth rate of new firms. The corresponding Pareto approximation to the Yule distribution for large i is $f(i) = \rho\Gamma(\rho + 1)i^{-(\rho+1)}$, where Γ is the gamma function (see (25) in ch. 4). Because of the exponential factor in the log series distribution, it will behave asymptotically in the upper tail like the geometric distribution, converging more rapidly toward zero, for $\rho > 1$, than will the Yule or Pareto distributions.

The log series and Yule distributions are compared in more detail in the introduction to ch. 2, where it is shown that the no-growth economy would have, in equilibrium, a smaller absolute number of firms than the growing economy of the same total size, and that there would be *relatively* fewer firms of very small and very large sizes, but relatively more medium-sized firms in the former than in the latter. Thus, there would be a lower degree of concentration, as it is usually measured, in the static than in the growing economy, as well as a concavity of the cumulative distribution, plotted on a log–log scale, to the origin.

The fact that concentration would be less in the static than in the growing economy, in spite of the operation of the Gibrat assumption of growth proportionate to size, appears at first to be paradoxical. The paradox disappears when we recall that in the static economy firms are losing sales according to the Gibrat assumption as well as gaining them, whereas in models having only a birth process and no death process, firms can only grow.

2.5. Combining firm size distributions

We conclude this introduction with a brief discussion of the relation between the size distribution of firms in an economy and the size distribution in its component industries. In particular, if the size distribution of firms in each industry is Pareto, will the size distribution of firms in the economy also be Pareto? What are the conditions under which Pareto distributions are "additive"?

Let us suppose that there are n nonoverlapping populations of firms, and that the size distribution of each population is Pareto, namely, $\log s + \beta_i \log r_i = M_i$, a constant. Here s is size and r_i rank, i.e. the number of firms in the ith population whose size is s or greater. Putting $1/\beta_i = \rho_i$, we can write, $r_i = (M_i/s)^{\rho_i}$, $i = 1, \ldots, n$. (See footnote 1 in ch. 10 for this relationship between β and ρ.)

For the total population of firms, we have $r = \Sigma\, r_i = \Sigma\, (M_i/s)^{\rho_i}$. In order for this to be Pareto, clearly it is necessary and sufficient to have the slope parameter ρ constant, where $\rho = -(\mathrm{d} \log r)/(\mathrm{d} \log s)$. From the definition of r, we have $\mathrm{d}r/\mathrm{d}s = -\Sigma\, \rho_i M_i^{\rho_i} s^{-\rho_i - 1} = -\Sigma\, \rho_i r_i/s$, hence $\rho = -(\mathrm{d} \log r)/(\mathrm{d} \log s) = -(s\,\mathrm{d}r)/(r\,\mathrm{d}s) = \Sigma\, \rho_i r_i/r$, which is the mean of ρ_i weighted by r_i/r. However, if not all ρ_i's are equal, for any i, r_i/r changes as s changes, hence ρ is not constant; while if all ρ_i's are equal, clearly ρ is constant. Thus, for the total population to be Pareto, the equality of all ρ_i's is the necessary and sufficient condition.

An interesting point to observe is that if not all ρ_i's are equal, the pooled distribution is strictly convex toward the origin throughout the entire range of s when $\log s$ is plotted against $\log r$. To show this;

$$(\mathrm{d}^2 \log r)/(\mathrm{d} \log s)^2 = \mathrm{d}\left(-\Sigma\, \rho_i r_i/r/(\mathrm{d} \log s)\right)$$

$$= -s \sum \rho_i (\mathrm{d}r_i r^{-1}/\mathrm{d}s)$$

$$= -s \sum \rho_i \left[-\rho_i r_i r^{-1}/s - (r^{-2})\left(-\sum \rho_i r_i/s\right) r_i\right]$$

$$= \sum \rho_i \left[\rho_i r_i / r - r_i \sum \rho_i r_i / r^2 \right]$$

$$= \sum \rho_i^2 r_i / r - \left(\sum \rho_i r_i / r \right)^2 > 0,$$

since the last expression is identical to that of the variance of a random variable whose realization ρ_i occurs with probability r_i / r.

We conclude that when two or more Pareto distributions are pooled together, the resulting distribution is Pareto if and only if all the distributions have identical slopes. What is more interesting is that if the pooled distribution is not Pareto, it is always convex to the origin regardless of the slopes of the individual component distributions.

This result is important in dealing with the aggregation of empirical firm-size distributions. For example, if the distributions for the individual industries as well as the pooled distribution for the economy are Pareto, then the slope parameters for the various industries must be nearly equal. Further, if the pooled distribution is concave to the origin, as is usually the case with the observed distributions, then it cannot arise as a pooling of Pareto distributions with different slopes.

With these introductory remarks, we turn now to our examination of stochastic models capable of generating skew distribution functions like those actually observed.

I

Skew distribution functions

CHAPTER 1

On a class of skew distribution functions*

This chapter, first published in 1955, provides the foundation for the series of investigations of skew distribution functions that is reported in this volume. The chapter sets out some simple axioms that generate the *Yule* distribution, which can be approximated in its upper tail by the *Pareto* distribution. It had already been observed independently by a number of investigators that one or the other of these distribution functions provided a good fit to the distributions of some disparate social phenomena. The aim of the research reported in the chapter was to determine if a relatively simple stochastic model could be found that would yield a distribution like these, and then to see if the assumptions of the model could be given plausible interpretations in terms of social processes. Results similar but not identical to those reported in this chapter had been reached by Yule (1924), Champernowne (1953) and Mandelbrot (1953).

The key axiom for such a stochastic model turns out to be what is usually called the *Gibrat* assumption. Roughly stated, this is the assumption that the expected percentage rate at which something will grow is independent of the size it has already attained. This assumption does not, by itself, determine what equilibrium distribution, if any, a system will reach, for the equilibrium depends also upon boundary conditions. Nevertheless, the Gibrat assumption virtually guarantees that the distribution will be highly skewed, with a long upper tail. This is the

*I have had the benefit of helpful comments from Messrs. Benoit Mandelbrot, Robert Solow and C. B. Winsten. I am grateful to the Ford Foundation for a grant-in-aid that made the completion of this work possible. [H.A.S.]

most striking characteristic of firm size distributions, with which this book is especially concerned. The terms "Gibrat assumption" and "Gibrat Law" will always be used here to mean the assumption that the expected percentage growth rate is independent of attained size. The log normal distribution is sometimes called the "Gibrat Law" or the "Gibrat distribution," but we shall refrain from using those names for the log normal distribution in order to avoid confusion.

Throughout the volume, only a very weak form of the Gibrat assumption is used. It is not assumed that the proportionality of expected growth to attained size holds for each unit in the population (e.g. each business firm), but only that it holds in the aggregate for all of the units in a given size class. Using this weak form of the assumption of course makes the mechanisms underlying the models much more plausible than if a stronger form were used.

As further tests of plausibility, this chapter and later ones also give attention to the sensitivity or insensitivity of the equilibrium distributions to small perturbations of the assumptions of the stochastic model. An encouraging characteristic of the skew distribution functions that we consider is their sturdiness – their relative stability in the face of such perturbations. This sturdiness gives us additional confidence that the models provide at least a good first-approximation explanation of the data.

The mathematical derivations in this and other chapters may be described as heuristic. We do not apologize for that, for our procedure lies squarely within the historical tradition of applied mathematics. The basic simplification we introduce in order to derive steady-state equilibria from the stochastic equations is to replace probabilities by their expected values. The conditions under which it is legitimate or illegitimate to do this have been thoroughly explored in the standard literature of stochastic processes. The reader who wishes to pursue the mathematics further might well begin his study with ch. 17 of Feller (1968) and the literature cited there. As further checks on the validity of our conclusions, particularly in cases where analytic solutions of the equations could not be obtained, we have, in later chapters, carried out a considerable number of simulations of variants of the basic stochastic process (see especially chs. 5 and 8).

1. Introduction

It is the purpose of this chapter to analyze a class of distribution functions that appears in a wide range of empirical data – particularly data describing sociological, biological and economic phenomena. Its appearance is so frequent, and the phenomena in which it appears so diverse, that one is led to the conjecture that if these phenomena have any property in common it can only be a similarity in the structure of the underlying probability mechanisms. The empirical distributions to which we shall refer specifically are: (1) distributions of words in prose samples by their frequency of occurrence, (2) distributions of scientists by number of papers published, (3) distributions of cities by population, (4) distributions of incomes by size, and (5) distributions of biological genera by number of species.

No one supposes that there is any connection between horse-kicks suffered by soldiers in the German army and blood cells on a microscope slide other than that the same urn scheme provides a satisfactory abstract model of both phenomena. It is in the same direction that we shall look for an explanation of the observed close similarities among the five classes of distributions listed above.

The observed distributions have the following characteristics in common:

(a) They are J-shaped, or at least highly skewed, with very long upper tails. The tails can generally be approximated closely by a function of the form

$$f(i) = (a/i^\kappa)b^i, \tag{1}$$

where a, b, and κ are constants; and where b is so close to unity that in first approximation the final factor has a significant effect on $f(i)$ only for very large values of i. Thus, for example, the number of words that occur exactly i times in James Joyce's *Ulysses* is about a/i^κ; the number of authors who published exactly i papers in *Econometrica* over a twenty-year period is approximately a/i^κ; and so on.

(*b*) the exponent, κ, is greater than 1, and in the cases of word frequencies, publication, and urban populations is very close to 2.[1]

(*c*) In the cases of word frequencies, publications and biological genera, the function (1) describes the distribution not merely in the tail but also for small values of i. In these cases the ratio $f(2)/f(1)$ is generally in the neighbourhood of one-third, and almost never reaches one-half; while $f(1)$ is generally in the neighbourhood of one-half.

Property (*a*) is characteristic of the "contagious" distributions – for example, the negative binomial as it approaches its limiting form, Fisher's logarithmic series distribution. However, in the case of the negative binomial, κ cannot exceed unity (and equals unity only in the limiting case of the log series); and if the distribution has a long tail, so that the convergence factor, b, is close to unity, $f(2)/f(1)$ cannot be less than one-half. Hence the negative binomial and Fisher's log series distributions do not provide a satisfactory fit for data possessing property (*a*) together with either (*b*) or (*c*).[2]

It is well known that the negative binomial and the log series distributions can be obtained as the stationary solutions of certain stochastic processes. For example, J. H. Darwin (1953) derives these from birth and death processes, with appropriate assumptions as to the birth- and death-rates and the initial conditions. In this paper we shall show that stochastic processes closely similar to those yielding the negative binomial or log series distributions lead to a class of functions having the three

---

[1]See Zipf (1949) for numerous examples of distributions with this property.
[2]The contrasting characteristics of distributions for which the log series provides a satisfactory fit and those, under consideration here, for which it does not are illustrated by examples (i) and (ii), respectively, in Good (1953).

properties enumerated above. This class of functions is given by

$$f(i) = AB(i, \rho + 1), \tag{2}$$

where A and ρ are constants, and $B(i, \rho + 1)$ is the beta function of $i, \rho + 1$:

$$B(i, \rho + 1) = \int_0^1 \tau^{i-1}(1 - \tau)^\rho \, d\tau$$

$$= \frac{\Gamma(i)\Gamma(\rho + 1)}{\Gamma(i + \rho + 1)}, \qquad 0 < i, 0 < \rho < \infty. \tag{3}$$

Now it is a well-known property of the gamma function (Titchmarsh, 1939, p. 58) that as $i \to \infty$, and for any constant κ,

$$\frac{\Gamma(i)}{\Gamma(i + \kappa)} \sim i^{-\kappa}. \tag{4}$$

Hence, from (3), we have, as $i \to \infty$:

$$f(i) \sim A\Gamma(\rho + 1)i^{-(\rho+1)}. \tag{5}$$

Therefore, the distribution (2) approximates the distribution (1) in the tail (more precisely, through the range in which the convergence factor of the latter is close to one). Further, if ρ is positive, κ will be greater than 1, as required by (b); and if ρ is equal to 1, κ will be equal to 2. It is easy to see that in the latter case we will have

$$f(i) = \frac{1}{i(i + 1)}, \quad \sum_{i=1}^{\infty} f(i) = 1, \tag{6}$$

so that $f(2)/f(1) = 1/3$; and $f(1) = 1/2$, as required by (c).

In the remainder of this paper we propose: (a) to describe a stochastic process that leads to the stationary distribution (2); (b) to discuss some generalizations of this process; and (c) to construct hypotheses as to why the empirical phenomena mentioned above can be represented, approximately, by processes of this general kind. Before proceeding, we should like to mention two earlier derivations, one of (2), the other of (1), that we have been able to discover in the literature.

Some thirty years ago, G. Udny Yule (1924) constructed a probability model, with (2) as its limiting distribution, to explain the distribution of biological genera by numbers of species. He also derived a modified form of (2), replacing the complete beta-function of (3) by the incomplete beta-function with upper limit of integration $\theta < 1$. (This modification has the same effect as the introduction of the convergence factor, b^i, in (1) – it causes a more rapid decrease in $f(i)$ for very large values of i [cf. Darwin (1953, p. 378); see also ch. 3, §7, below].) It seems highly appropriate to call the distribution (2) the Yule distribution.

Because Yule's paper predated the modern theory of stochastic processes, his derivation was necessarily more involved than the one we shall employ here. Moreover, while the assumptions he required are plausible for the particular biological problem he treated, the corresponding assumptions applied to the four other phenomena we have mentioned appear much less plausible. Our derivation requires substantially weaker assumptions than Yule's about the underlying probability mechanism.

More recently D. G. Champernowne (1953) has constructed a stochastic model of income distribution that leads to (1) and to generalizations of that function. Since the points of similarity between his model and the one under discussion here are not entirely obvious at a first examination, we shall consider their relation in a later section of this chapter.

2. The stochastic model

For ease of exposition, the model will be described in terms of word frequencies. In a later section, alternative interpretations will be provided. Our present interest is in the kind of stochastic process that would lead to (2).

Consider a book that is being written, and that has reached a length of k words. We designate by $f(i, k)$ the number of *different* words that have occurred exactly i times in the first k words. That is, if there are 407 different words that have occurred exactly once each, then $f(1, k) = 407$.

Assumption I. The probability that the $(k + 1)$-st word is a

word that has already appeared exactly i times is proportional to
if(i, k) – that is, to the total number of occurrences of all the
words that have appeared exactly i times.

Note that this assumption is much weaker than the assumption (I′) that the probability a *particular* word occur next be proportional to the number of its previous occurrences. Assumption (I′) implies (I), but the converse is not true. Hence we leave open the possibility that, among all words that have appeared i times the probability of recurrence of some may be much higher than of others.

Assumption II. There is a constant probability, α, that the $(k + 1)$-st word be a new word – a word that has not occurred in the first k words.

Assumptions (I) and (II) describe a stochastic process, in which the probability that a particular word will be the next one written depends on what words have been written previously. If this process correctly describes the selection of words, then the words in a book cannot be regarded as a random sample drawn from a population with a prior distribution. The reasonableness of the former, as compared with the latter type of explanation of the observed distributions, will be discussed in §4.

From (I), it follows that

$$E\{f(i, k + 1)\} - f(i, k) = U(k)\{(i - 1)f(i - 1, k) \\ - if(i, k)\}, \quad i = 2, \ldots, k + 1,$$
(7)

where E is an expectation operator and $U(k)$ is a normalizing factor, for if the $(k + 1)$st word is one that has previously occurred $(i - 1)$ times, $f(i, k + 1)$ will be increased over $f(i, k)$, and the probability of this, by assumption (I), is proportional to $(i - 1)f(i - 1, k)$; if the $(k + 1)$st word is one that previously occurred i times, $f(i, k + 1)$ will be decreased, and the probability of this, by assumption (I), is proportional to $if(i, k)$; while in all other cases, $f(i, k + 1) = f(i, k)$.

From (I) and (II) we obtain similarly

$$E\{f(1, k + 1)\} - f(1, k) = \alpha - U(k)f(1, k),$$

$$0 < \alpha < 1. \quad (8)$$

Since we will be concerned throughout with "steady-state" distributions (as defined by equation (14) below), we replace the expected values in (7) and (8) by the actual frequencies. (Alternatively, we might replace frequencies on the right-hand side of the equation by probabilities.) That is, we write, instead of (7) and (8),

$$f(i, k + 1) - f(i, k) = U(k)\{(i - 1)f(i - 1, k) - if(i, k)\},$$

$$i = 2, \ldots, k + 1, \quad (9)$$

$$f(1, k + 1) - f(1, k) = \alpha - U(k)f(1, k), \quad (10)$$

where the f's now represent expected values.

Now, we wish to evaluate the factor of proportionality $U(k)$. Since $U(k)if(i, k)$ is the probability that the $(k + 1)$st word is one that previously occurred i times, we must have

$$\sum_{i=1}^{k} U(k)if(i, k) = U(k) \sum_{i=1}^{k} if(i, k) = 1 - \alpha. \quad (11)$$

But $\sum_{i=1}^{k} if(i, k)$ is the total number of words up to the kth, hence

$$\sum_{i=1}^{k} if(i, k) = k, \quad (12)$$

and

$$U(k) = \frac{1 - \alpha}{k}. \quad (13)$$

Substituting (13) in (9) and (10), we could solve these differential equations explicitly. We can avail ourselves, however, of a simpler – though nonrigorous – method for discovering the solutions, and can then test their correctness by substitution in the original equations. Consider the "steady-state" distribution in the following sense. We assume

$$\frac{f(i, k + 1)}{f(i, k)} = \frac{k + 1}{k}, \qquad \text{for all } i \text{ and } k; \qquad (14)$$

so that all the frequencies grow proportionately with k, and hence maintain the same relative size. (Since we must have $f(i, k) = 0$ for $i > k$, equation (14) cannot hold exactly for all i and k. But as explained above, we are concerned at the moment with heuristic rather than proof.)

From (14) it follows that

$$\frac{f(i, k)}{f(i - 1, k)} = \frac{f(i, k + 1)}{f(i - 1, k + 1)} = v(i), \qquad (15)$$

where $v(i)$ does not involve k. Hence, the *relative* frequencies, which we will designate by $f(i)$, are independent of k. Substituting (13), (14) and (15) in (9), we get

$$\left(\frac{k + 1}{k} - 1\right) f(i, k) = \frac{(1 - \alpha)}{k} \left\{\frac{(i - 1)}{v(i)} - i\right\} f(i, k). \qquad (16)$$

Canceling the common factor, and solving for $v(i)$, we obtain

$$v(i) = \frac{(1 - \alpha)(i - 1)}{1 + (1 - \alpha)i} = \frac{f(i)}{f(i - 1)}, \qquad i = 2, \ldots, k. \qquad (17)$$

For convenience, we introduce

$$\rho = \frac{1}{1 - \alpha}, \qquad 1 < \rho < \infty. \qquad (18)$$

Since $f(i) = v(i)f(i - 1) = v(i) \cdot v(i - 1) \cdots v(2)f(1)$, we obtain from (17) and (18),

$$f(i) = \frac{(i - 1)(i - 2) \cdots 2 \cdot 1}{(i + \rho)(i + \rho - 1) \cdots (2 + \rho)} f(1)$$

$$= \frac{\Gamma(i)\Gamma(\rho + 2)}{\Gamma(i + \rho + 1)} f(1)$$

$$= (1 + \rho)B(i, \rho + 1)f(1), \qquad i = 2, \ldots, k. \qquad (19)$$

The second relation follows from the fact that

$$\Gamma(i + \rho + 1) = (i + \rho)\Gamma(i + \rho)$$
$$= (i + \rho)(i + \rho - 1) \cdots (2 + \rho)\Gamma(\rho + 2).$$

(20)

But (19) is identical with (2) if we take $A = f(1)(1 + \rho)$. We shall later show that A is indeed equal to ρ (see ch. 3).

That (19) is in fact a solution of (9) can be verified by direct substitution. Moreover, it is in the following sense a stable solution. Suppose that (17) is *not* satisfied. Whatever be the values of the $f(i, k)$ for a given k, we may write without loss of generality

$$\frac{f(i, k)}{f(i - 1, k)} = \frac{(1 - \alpha)(i - 1)}{(1 - \alpha)i + 1 + \varepsilon(i, k)},$$

(21)

where $\varepsilon(i, k)$ is some function of i and k. If we now divide both sides of (9) by $f(i, k)$ and substitute (21) in the right-hand side of the resulting equation, we obtain after simplification,

$$\frac{f(i, k + 1)}{f(i, k)} = \frac{k + 1 + \varepsilon(i, k)}{k}.$$

(22)

Hence the ratio of $f(i, k + 1)$ to $f(i, k)$ will be greater than $(k + 1)/k$ if $\varepsilon(i, k)$ is positive, and less than $(k + 1)/k$ if $\varepsilon(i, k)$ is negative. Since new words are introduced at a constant rate, $\sum_1^k f(i, k)$ must be proportional to k; therefore, by (22), we have

$$\sum_{i=1}^{k+1} f(i, k + 1) - \frac{k + 1}{k} \sum_{i=1}^{k+1} f(i, k) = \frac{1}{k} \sum_{i=1}^{k} \varepsilon(i, k)f(i, k) = 0.$$

(23)

We may interpret the three equations, (21)–(23), as follows. In an average sense, the frequencies will grow proportionately with k. If a particular frequency is "too large" compared with the next lower frequency ($\varepsilon(i, k)$ negative in (21)), it will grow at a rate slower than the average; if it is "too small" ($\varepsilon(i, k)$ positive), it will grow more rapidly than the average.

It remains to be shown that $f(i) = B(i, \rho + 1)f(1)$ is a proper distribution function. In particular, we require that $\Sigma_{i=1}^{k} iB(i, \rho + 1)$ converge as $k \to \infty$. Now, it is well known that $\Sigma_{i=1}^{\infty} i^{-a}$ converges for every $a > 1$. But by (4),

$$\sum_{i=1}^{\infty} iB(i, \rho + 1) \sim i \cdot i^{-(\rho+1)} = i^{-\rho}. \tag{24}$$

Hence, by the usual ratio comparison test, $\Sigma_{i=1}^{\infty} iB(i, \rho + 1)$ converges for $\rho > 1$, as required.

From the definition of α the total number, n_k, of *different* words will be αk; while the total number of word occurrences is k. That is,

$$n_k = \sum_{i=1}^{k} f(i, k) = \alpha k = \alpha \sum_{i=1}^{k} if(i, k). \tag{25}$$

Returning to (10), and using (14), we get

$$\left(\frac{k+1}{k} - 1\right) f(1, k) = \alpha - \frac{1-\alpha}{k} f(1, k), \tag{26}$$

whence

$$f(1, k) = \frac{k\alpha}{2 - \alpha} = \frac{n_k}{2 - \alpha}, \tag{27}$$

or $f(1) = f(1, k)/n_k = 1/(2 - \alpha)$.

From (18) and (27), and by successive application of (17), we can compute the values of ρ, $f(1)/n_k$, $f(2)/n_k$, $f(3)/n_k$, etc., for given values of α (Table 1.1).

Table 1.1

α	ρ	$f(1)/n_k$	$f(2)/n_k$	$f(3)/n_k$
0.0	1	0.500	0.167	0.083
0.1	1.11	0.527	0.169	0.082
0.2	1.25	0.556	0.171	0.080
0.3	1.43	0.588	0.171	0.077
0.5	2.00	0.667	0.167	0.067
0.7	3.33	0.769	0.144	0.046
0.9	10.00	0.909	0.076	0.012

Thus far we have considered the case where α, the rate at which new words are introduced, is independent of k. We can easily generalize to the case where α is a function of k by making the appropriate substitution in (10). The equations can then be solved directly, but the method employed to obtain a "steady-state" distribution is not applicable, since it is not easy to define what is meant by the steady state in this more general case. We will content ourselves with some approximate results for two special cases. These special cases will give us insight as to how a distribution function may arise which, for small values of i, can be approximated by (2), with $0 < \rho < 1$.

Case I. Suppose the system to be in the steady state described by (14) with $k = k_0$, and that the flow of *new* words suddenly ceases, so that $\alpha(k) = 0$ for $k > k_0$. We will now have $U(k) = 1/k$ for $k > k_0$, and (10) becomes

$$f(1, k + 1) = \left(1 - \frac{1}{k}\right) f(1, k) = \frac{k - 1}{k} f(1, k). \qquad (28)$$

We define

$$z(i) = \frac{f(i, k + 1)}{f(i, k)}, \qquad i = 2, \ldots, k + 1. \qquad (29)$$

Since no new words are being introduced, we must have

$$n_k = f(1, k) + \sum_{i=2}^{k} f(i, k) = f(1, k + 1) + \sum_{i=2}^{k} f(i, k + 1)$$

$$= \frac{(k - 1)}{k} f(1, k) + \sum_{i=2}^{k} z(i) f(i, k), \qquad (30)$$

whence

$$\frac{\sum_{i=2}^{k} [z(i) - 1] f(i, k)}{\sum_{i=2}^{k} f(i, k)} = \frac{1}{k} \frac{f(1, k)}{\sum_{i=2}^{k} f(i, k)}. \qquad (31)$$

Let us define ρ_i using $v(i)$ in (15), namely,

$$v(i) = \frac{f(i, k)}{f(i - 1, k)} = \frac{(i - 1)}{(1 + \rho_i)} \qquad (32)$$

(where we suppose that ρ_i changes only slowly with k). Instead of (9), we have

$$f(i, k+1) - f(i, k) = \frac{1}{k}[(i-1)f(i-1, k) - if(i, k)].$$

(33)

Substituting (29) and (32) in this, we get

$$z(i) - 1 = \frac{1}{k}[(i + \rho_i) - i],$$ (34)

whence

$$\rho_i = k(z(i) - 1),$$ (35)

and

$$\bar{\rho} = \frac{\sum_{i=2}^{k} k(z(i)-1)f(i, k)}{\sum_{i=2}^{k} f(i, k)} = \frac{f(1, k)}{\sum_{i=2}^{k} f(i, k)} = \frac{f(1, k)}{n_k - f(1, k)}.$$

(36)

Define

$$\hat{\alpha} = f(1, k)/n_k;$$ (37)

Then

$$\bar{\rho} = \frac{\hat{\alpha}}{1 - \hat{\alpha}} \quad \text{and} \quad 0 < \bar{\rho} < \infty.$$ (38)

Proceeding heuristically, we can see that after α becomes zero, $f(1, k)$ will begin to decrease with k, and the value of ρ_i will be larger the larger is i. For small values of i, we will have $\rho_i < \bar{\rho}$, and for large values $\rho_i > \bar{\rho}$. However, the tail of the distribution will be affected only slowly by the change in α. Hence, we may suppose that $\lim_{i \to k_0} (\rho_i) = \rho_0$, where ρ_0 is ρ_{k_0}. On the other hand, since the weighted average in (35) is heavily influenced by the large frequencies for small values of i, ρ_i will be only slightly less than $\bar{\rho}$. Hence we may expect the distribution to take the form of a slightly curved line on a double-log scale, with a slope of

$-(\bar{\rho}+1)$ at the lower end, and a slope of $-(\rho_0+1)$ at the upper end. If $\rho_0 > 2$, then Σ $if(i, k)$ will converge. An example of such a distribution will be given in §4.

Case II. A second approximate solution can be obtained if we assume that α decreases with k, but very slowly. By definition, we have $\alpha = dn_k/dk = n'$. The condition for a steady state (all frequencies increasing proportionately) is now

$$f(i, k + 1) = [1 + (n'/n_k)]f(i, k).\tag{39}$$

Substituting as before, (13) and (39) in (9), we again obtain (19), where ρ is now given by

$$\rho = \frac{n'k}{n_k}\frac{1}{(1-n')}.\tag{40}$$

The slope obtained in the derivation for constant α has now been multiplied by the factor $(n'k)/n_k$, which for monotonically decreasing α is less than one. Hence, the effect of a decrease in the rate of introduction of new words is to lengthen the tail of the distribution, as was also true in case I. If the new value of ρ is less than one, we do not have a proper distribution function (see equation (24)), hence the equation can hold only for small and moderate values of i, and there must be a curve (on a logarithmic scale) in the tail of the distribution.

3. An alternative formulation of the process

There are some alternative ways for deriving the relation (19). One of these will be useful to us when we come, in the next section, to a more specific discussion of word frequencies and frequencies of publications. Moreover, this derivation avoids the difficulties we have encountered in the definition of "steady state."

Equation (17) may be written

$$0 = (1 - \alpha)[(i - 1)f(i - 1) - if(i)] - f(i),$$
$$i = 2, \ldots, k,\tag{41}$$

where we have substituted $f(i)$ for $f(i, k)$.

Similarly, from (10), we obtain

$$0 = 1 - (1 - \alpha)f(1) - f(1). \tag{42}$$

These two equations may be interpreted as follows. We consider a sequence of k words. We add words to the sequence in accordance with assumptions (I) and (II) of §2, *but we drop words from the sequence at the same average rate*, so that the length of the sequence remains k. The method according to which we drop words is the following:

Assumption III. If one representative of a particular word is dropped, then all representatives of that word are dropped, and the probability that the next word dropped be one with exactly i representatives is α times $f(i)$.

This assumption would be approximately satisfied, for example, if the representatives of each word, instead of being distributed randomly through the sequence, were closely "bunched." This possibility is consistent with assumption (I).

Eq. (41), in our new interpretation, may be regarded as the steady-state equilibrium of the stochastic process defined by

$$k[f(i, m + 1) - f(i, m)] = (1 - \alpha)[(i - 1)f(i - 1, m)$$
$$- if(i, m)] - f(i, m), \tag{43}$$

where m is now not the total number of words (which remains a constant, k), but the number of additions to (and withdrawals from) an initial arbitrary sequence of k words. Since the k of this process, unlike that of §2, remains constant, the ordinary proofs of the existence of a unique steady-state solution will apply (see Feller (1950, p. 373)), and we avoid the troublesome questions of rigor that confronted us in §2.

The solution of (41) and (42) is, of course, again given by

$$\frac{f(i)}{f(i - 1)} = \frac{(1 - \alpha)(i - 1)}{1 + (1 - \alpha)i}. \tag{17}$$

If we were to replace the last terms of (41) and of (42),

respectively, by terms corresponding to the usual form of the death process, we would have (cf. Darwin (1953, p. 375) and Kendall (1948))

$$0 = (1 - \alpha)[(i - 1)f(i - 1) - if(i)]$$
$$- [if(i) - (i + 1)f(i + 1)], \quad i = 2, \ldots, k - 1, \qquad (44)$$

$$0 = 1 - (1 - \alpha)f(1) - [f(1) - 2f(2)]. \qquad (45)$$

The solution of this system of equations is

$$\frac{f(i)}{f(i - 1)} = \frac{(1 - \alpha)(i - 1)}{i}, \qquad (46)$$

which is Fisher's logarithmic series distribution.

Since the log series distribution is a limiting case of the negative binomial, we may ask whether there is a distribution that stands in the same relation to the latter as (17) stands in relation to (46). We can obtain such a distribution by a modification of the birth process in (41). We assume now that the birthrate is the sum of two components – one proportional to $if(i)$, the other proportional to $f(i)$. In place of (41) we have

$$0 = \frac{(1 - \alpha)k}{k + \chi}[(i - 1 + \chi)f(i - 1) - (i + \chi)f(i)] - f(i), \qquad (47)$$

where χ is a constant of proportionality for the second component, the solution of which is

$$\frac{f(i)}{f(i - 1)} = \frac{\lambda(i - 1 + \chi)}{\lambda(i + \chi) + 1} = \frac{(i - 1 + \chi)}{(i + \chi + 1/\lambda)}, \qquad (48)$$

where $\lambda = k(1 - \alpha)/(k + \chi)$.

A rather remarkable property of (48) is that in the tail it still has the limiting form (1) with $b = 1$. Hence for α and χ small, this generalized Yule distribution will still possess the three properties listed in the introduction. The fact that a reasonably wide range of variation in the assumptions underlying the stochastic model does not alter greatly the form of the distribution adds plausibility to the use of such stochastic processes to explain the observed distributions. Our next task is to consider these explanations in more detail.

4. The empirical distributions

In this section we shall try to give theoretical justifications for the observed fit of the Yule distribution to a number of different sets of empirical data.

4.1. Word frequencies

A substantial number of word counts have been made, in English and in other languages (see Hanley (1937), Thorndike (1937), Yule (1944), Zipf (1949) and Good (1953)). Eq. (6) provides a good fit to almost all of them. When the more general function, (2), is used, the estimated value of ρ is always close to 1. When a convergence factor, b^i, is introduced to account for the deficiency in frequencies for very large values of i, the estimated value of b is also very close to 1. Good (1953), for instance, applies (6) multiplied by a convergence factor to the Eldridge count, and obtains $b = 0.999667$.

These regularities are the more surprising in that the various counts refer to a quite heterogeneous set of objects. In the Yule and Thorndike counts, inflected forms are counted with the root word; in most of the other counts each form is regarded as a distinct word. The Yule counts include only nouns; the others, all parts of speech. The Dewey, Eldridge and Thorndike counts are composite – compiled from a large number of separate writings; most of the others are based on a single piece of continuous prose. We would regard this heterogeneity as further evidence that the explanation is to be sought in a probability mechanism, rather than in more specific properties of language; but at the same time, the heterogeneity complicates the task of specifying the probability mechanism in detail. We shall avoid

questions of "fine structure" – which would require an expert-
ness in linguistics we do not possess – and confine ourselves to
three broad problems: (1) the distribution of word frequencies in
the whole historical sequence of words that constitutes a language: (2) the distribution of word frequencies in a continuous
piece of prose: (3) the distribution of word frequencies in a
sample of prose assembled from composite sources.

(1) For obvious reasons, we do not have any empirical data on
the cumulated word frequencies for a whole language. On *a
priori* grounds, it does not appear unreasonable to postulate that
these frequencies are determined by a process like that described in §2. The parameter α is then the rate at which
neologisms appear in the language as a fraction of all word
occurrences – and hence α can be assumed to be very close to
zero.

(2) The process of §2 might also describe the growth of a
continuous piece of prose – for example, Joyce's *Ulysses*. But
there are some serious objections to this hypothesis. An author
writes not only by processes of *association* – i.e. sampling
earlier segments of the word sequence – but also by processes of
imitation – i.e. sampling segments of word sequences from other
works he has written, from works of other authors, and, of
course, from sequences he has heard. The model of §2 apparently allows only for association, and excludes imitation.

The word frequencies in *Ulysses* provide obvious evidence of
the importance of both processes. The fact that the proper noun
"Bloom" occurs 926 times and ranks 30th in frequency must be
attributed to association. If Joyce had named his hero "Smith,"
that noun, instead of "Bloom," would have ranked 30th. On the
other hand, "they," which occurs 1010 times in *Ulysses* and
ranks 27th, has very nearly the same rank – the 28th – in the
Dewey count. In fact, of the 100 most frequent words in *Ulysses*,
78 are among the top 100 in the Dewey count. This similarity in
ranking of "common" words argues for imitation rather than
association. Even for the common words, however, the varia-

tions in frequency from one count to another are far too great to be explained as fluctuations resulting from random sampling from a common population of words. The imitative process must involve stratified sampling, and imitation must be compounded with association.

It is worth emphasizing again at this point that assumption (I) does not require that the choice of the next word from among those previously written be completely random. Suppose, for example, that a writer were to assign to each page he has already written a number, p_j, $\Sigma p_j = (1 - \alpha)$, the size of p_j varying with the "affinity" of the subject discussed on the jth page to the subject next to be discussed. If his next word were selected by a stratified sampling of the previous pages, with probability p_j for each page, then eq. (7) would generally be satisfied. For although individual words would be distributed unevenly through the preceding pages, the totality of words having a given frequency, i, in all the previous pages taken together would be distributed almost evenly through these pages. Hence, the various frequency strata would have proportionate probabilities of being sampled, for most choices of the p_j. That is all that is required for eq. (7). This same comment applies to the assumption we shall subsequently make regarding imitative sampling from other works.

Let us now reconsider the problem of a piece of continuous prose. Since both the processes of association and imitation are involved, the sequence that is counted is to be regarded as a "slice," of length k, of the entire sequence of words in the language, or of the entire sequence written by the author. Hence the word count is better described by the stochastic process of §3 than by the process of §2.

In determining the probability that a word selected in such a sequence be one that has occurred exactly i times, we must consider separately the process of imitation and association. Assume that, on the average, a fraction p_m, of the words added is selected by imitation, and the remaining fraction, $(1 - p_m)$, by association. Since no new words can be introduced by association, the joint probability that the next word will be selected by association and will be a word that has already occurred i times is $(1 - p_m)if(i, k)/k$.

The words selected by imitation present a more difficult problem, and we shall have to content ourselves with a reasonable assumption that has no rigorous justification. On the average, a word that has occurred i times will have a chance less than i/k of being the next one chosen by imitation, because in the sequence that is being sampled there are words that have not yet been chosen at all, and because with progressive change of subject, different strata of the language will be sampled. Since words with large i will generally be "common" words, fairly uniformly distributed through all strata of the language, the deficiency may be expected to be proportionately greater for small i than for large i. As a rough, but reasonable, approximation let us assume that: the joint probability that the next word will be selected by imitation and will be a word that has already occurred i times is $p_m(i - \omega)f(i, k)/k$, where $0 < \omega < 1$. (Our result would not be essentially altered if we wrote $\omega(i)$ instead of ω, provided only that $\omega(i)$ does not vary a great deal.)

Adding the two joint probabilities – for association and imitation, respectively – we find that the total probability that the next word be one that has occurred i times is $(i - p_m\omega)f(i, k)/k$. By summing this probability over i and subtracting from 1, we find that the probability that the next word be a new word is $p_m\omega(n_k/k)$.

If the method of dropping words from the sequence satisfies assumption (III) of §3, we set the difference between the birth-rate and the death-rate equal to zero, and obtain the steady-state equation

$$0 = (i - p_m\omega - 1)f(i - 1) - (i - p_m\omega)f(i) - f(i), \qquad (49)$$

which has as its solution

$$\frac{f(i)}{f(i - 1)} = \frac{(i - p_m\omega - 1)}{(i - p_m\omega + 1)}. \qquad (50)$$

Again, we obtain a distribution with the required properties.

(3) The distribution of word frequencies in a sample of prose assembled from composite sources can be explained along the same general lines. Again, we may regard the sample as a "slice" from a longer sequence, but we might expect the parameters ω and p_m to be somewhat larger than in a comparable piece of continuous prose. The qualification "comparable" is important, for ω may be expected to be smaller for homogeneous prose using a limited vocabulary of common words than for prose with a large vocabulary and treating of a variety of subjects. Hence ω might well be larger for the continuous *Ulysses* count than for the Eldridge count, which is drawn from newspaper sources. Indeed, the empirical evidence suggests that this is the case.

There is no point in elaborating the explanation further. What has been shown is that the observed frequencies can be fitted by distributions derived from probability assumptions that are not without plausibility.

A very different and very ingenious explanation of the observed word-frequency data has been advanced recently by Dr. Benoit Mandelbrot (1953). His derivation rests on the assumption that the frequencies are determined so as to maximize the number of bits of information, in the sense of Shannon, transmitted per symbol. There are several reasons why we prefer an explanation that employs averaging rather than maximizing assumptions. First, an assumption that word usage satisfies some criterion of efficiency appears to be much stronger than the probability assumptions required here. Secondly, numerous doubts, which we share, have been expressed as to the relevance of Shannon's information measure for the measurement of semantic information.

Before leaving the subject of word frequencies, it may be of interest to look at some of the empirical data. Good (1953, pp. 257–60), has obtained good fits to the Eldridge count and to one of Yule's counts by the use of eq. (6). Table 1.2 summarizes a few of the data on two word counts, and compares the actual frequencies, $f(1, k)$, $f(2, k)$ and $f(3, k)$ with the frequencies estimated from eq. (3). The actual values of k and n_k are used to estimate $\alpha = n_k/k$, and (17) and (27) to obtain the expected

Table 1.2

Word count	$\alpha = \dfrac{n_k}{k}$	$f(1, k)$		$f(2, k)$		$f(3, k)$	
		Actual	Estimate	Actual	Estimate	Actual	Estimate
Ulysses (Hanley (1937))	0.115	16,432	15,850	4,776	4,870	2,194	2,220
Eldridge (Good (1953))	0.136	2,976	3,220	1,079	977	516	400

frequencies. In both cases the observed value of n_k/k leads to an estimate of ρ in the neighborhood of 1.1 to 1.2. An empirical fit to the whole distribution of a function of the form $f(i, k) = C(k)i^{-(\rho+1)}$ gives an estimated value of ρ, in both cases, of about one – in reasonable agreements with the first estimate. A good fit to both the *Ulysses* and the Eldridge counts can also be obtained from (50), with ω equal to about 0.2 in the former case, and close to zero in the latter.

In the case of Thorndike's count of $4\frac{1}{2}$ million words in children's books (Thorndike (1937)), we may assume that the supply of new words was virtually exhausted before the end of the count. In his count $f(1, k)$ is substantially below $0.5 n_k$ (about $0.34 n_k$), as we would expect under these circumstances (see case I of §2). Thorndike estimated the empirical value of our $\bar{\rho}$ at 0.45, which is entirely consistent with the observed value of $0.34 n_k$ for $f(1, k)$. For, by (38), $\alpha = \bar{\rho}/(\bar{\rho} + 1) = 0.31$.

4.2. Scientific publications

At least four sets of data are available on the number, $f(i, k)$, of authors contributing a given number, i, of papers each to a journal or journals (Davis (1941), Leavens (1953)). These are counts of (a) papers written by members of the Chicago Section of the American Mathematical Society over a 25-year period; (b) papers listed in *Chemical Abstracts* (under A and B) over 10 years; (c) papers referred to in a history of physics; and (d) papers and abstracts in *Econometrica* over a 20-year period.

We may postulate a mechanism like that of §3, eq. (41). The authorship of the next paper to appear is "selected" by stratified

Table 1.3
Number of persons contributing.

No. of contributions	Chicago Math. Soc.		Chem. Abstracts		Physicists		Econometrica	
	Actual[a]	Estimate	Actual[a]	Estimate	Actual[a]	Estimate	Actual[b]	Estimate
1	133	—	3,991	4,050	784	824	436	453
2	43	46	1,059	1,160	204	217	107	119
3	24	23	493	522	127	94	61	51
4	12	14	287	288	50	50	40	27
5	11	10	184	179	33	30	14	16
6	14	7	131	120	28	20	23	11
7	5	5	113	86	19	14	6	7
8	3 ⎫	4 ⎫	85	64	19 ⎫	10 ⎫	11 ⎫	5 ⎫
9	9 ⎬ 13	3 ⎬ 10	64	49	6 ⎬ 32	8 ⎬ 24	1 ⎬ 12	4 ⎬ 12
10	1 ⎭	3 ⎭	65	38	7 ⎭	6 ⎭	0 ⎭	3 ⎭
11 or more	23	30	419	335	48	52	22	25
Estimated α	0[c]		0.30		0.39		0.41	
Estimated ρ	0.916		1.43		1.64		1.69	
k	1,124		22,939		3,396		1,759	
n_k	278		6,891		1,325		721	

[a]Davis (1941).
[b]Leavens (1953).
[c]$\rho = \bar{\rho}$ estimated in this case from (37) to (38).

sampling from the strata of authors who have previously pub-lished 1, 2, . . . , papers, the probability for each stratum being proportional to $if(i)$. Again, the probabilities for individual authors need not be proportional to i, but only the probabilities for the aggregates of authors with the same i. For example (as in the case of words), the probability for a particular author may be higher if he has published recently than if he has not. The gradual retirement of authors corresponds to assumption (III).

A comparison of the actual frequencies, for i from 1 to 10, with the estimated frequencies, derived from (17) and (27), is shown in table 1.3. The fit is reasonably good, when it is remembered that only one parameter is available for adjustment. However, it should be noted that the estimated frequencies tend to be too high for $i = 1, 2$ and too low for $i = 3, \ldots, 10$. In three of the four cases, they are again too high for the tails of the distributions. A further refinement of the model is apparently needed to remove these discrepancies.

4.3. City sizes

It has been observed, for every U.S. Census since the early nineteenth century, and for most other Western countries as well, that if $F(i)$ is the relative frequency of cities of population greater than or equal to i, namely $F(i) = \Sigma_{j=i}^{\infty} f(j)$, then, as derived in (24) in ch. 4,

$$F(i) \sim \Gamma(\rho + 1)i^{-\rho}, \tag{51}$$

where ρ is close to 1 (see Zipf (1949, chs. 9, 10)).

Again, we would expect such a distribution if the underlying mechanism were one describable by equations like (9) and (10). Such a mechanism is not hard to conceive. First, eq. (9) would hold if the growth of population were due solely to the net excess of births over deaths, and if this net growth were proportional to present population. This assumption is certainly satisfied at least roughly. Moreover, it need not hold for each city, but only for the aggregate of cities in each population band.

Finally, the equation would still be satisfied if there were net migration to or from cities of particular regions, provided the net addition or loss of population of individual cities *within any region* was proportional to city size. That is, even if all California cities were growing, and all New England cities declining, the equation would hold provided the percentage growth or decline in each area were uncorrelated with city size.

In the case of cities, eq. (51) could only be expected to hold down to some minimum city size – say, 5000 or 10,000. The constant α would then be interpreted as the fraction of the total population growth in cities above the minimum size that is accounted for by the new cities that reach that size.

4.4. Income distribution

Vilfredo Pareto is generally credited with the discovery that if personal incomes are ranked by size, the number of persons, $F(i)$, whose incomes exceed i can be approximated closely, for the upper ranges of income, by eq. (51) with ρ usually in the neighborhood of 1.5 (Davis (1941), Champernowne (1953)). Hence, the income distributions bear a family resemblance in their upper ranges to those we have already considered, although the parameter ρ, is substantially larger than 1 – its characteristic value in the case of word frequencies and city size distributions.

A stochastic mechanism similar to those described in §3 would again produce steady-state distributions closely resembling the observed ones. We picture the stream of income as a sequence of dollars allocated probabilistically to the recipients. If the total annual income of all persons above some specified minimum income is k dollars, the segment of this sequence running from the mth to the $(m + k)$th dollar is the income for the year beginning at time m. We assume that the probability that the next dollar will be allotted to some person with an annual income of i dollars is proportional to $(i + \chi)f(i)$, with χ positive but small. This represents a modification of assumption (I) that decreases the proportion of the total stream going to persons of high income relative to the proportion going to persons with incomes close to the minimum. We assume that a fraction of the dollars is assigned to new persons – i.e. persons reaching the minimum income to which the assumptions apply (assumption (II)). We assume that there is considerable variance among persons within each income class in the probability of receiving additional income, so that the rate at which dollars are dropped from any income class as m increases satisfies assumption (III). Then we obtain again eq. (48), which now holds for i greater than the minimum income. For large i, this distribution has the required properties with $1/\lambda = \rho$.

The same result has been reached by D. G. Champernowne (1953), following a somewhat different route. He divides income recipients at time t_1 into classes of equal proportionate width. That is, if I_0 is the minimum income considered, then the first class contains persons with incomes between I_0 and νI_0, the second class, persons with incomes between νI_0 and $\nu^2 I_0$, and so on. Next he introduces transition probabilities p_{jl}, that a person who is in class j at time t_1 will be in class l at time t_2. He assumes that p_{jl} is a function only of $(j - l)$. Now, by his definition of the income classes, the average income of persons in class j will be about $\nu^{(j-l)}$ times the average income of persons in class l. Hence, the expected income at t_2 of a person who was in class j at t_1 will be

$$\sum_l p_{jl} I_l = \sum_l p(j - l)\nu^{(j-l)} I_j = \alpha I_j \qquad (\alpha \text{ a constant}), \quad (52)$$

where I_j is the average income in class j. But Champernowne assumes explicitly that $\alpha < 1$. From this it is clear that his model satisfies our assumptions (I) (in its original form) and (II). Further, since he assumes a substantial variance in income expectations among persons in a given class, our assumption (III) is also approximately satisfied. Hence, in spite of the surface differences between his model and those developed here, the underlying structure is the same.

4.5. Biological species

We conclude this very incomplete list of phenomena exhibiting the Yule distribution by mentioning the example originally analysed by Yule himself (1924). It was discovered by Willis that the relative frequency, $f(i)$, of genera of plants having i species each was distributed approximately according to (51), with $\rho < 1$. Yule explained these data by a probability model in which the probability, p_s, of a specific mutation occurring in a particular genus during a short time interval was proportional to the number of species in the genus; while the probability, p_g, of a generic mutation during the same interval was proportional to the number of genera. Starting at t_0 with a single genus of one species, he computed the distribution $f(i, t)$ for $t_1, t_2, \ldots$, and found the limit as $t \to \infty$. This limiting distribution corresponds to (19) with $\rho = p_g/p_s$. Yule observed that for $p_g < p_s$ (as required to fit the empirical data), this was not a proper distribution function, and obtained the approximate distribution for $t = T$. His procedure was equivalent to replacing the complete beta function in (19) by the incomplete beta function, taking as the upper limit of integration an appropriate function of T.

If, in the process of §2, we define k as the total number of different species and $f(i, k)$ as the number of genera with exactly i species, we see that our k is a monotonic increasing function of Yule's t (specifically, $k = e^{p_s t}$). Making the appropriate transformation of variables, we find that Yule's assumption with respect to the rate of specific mutation corresponds to our assumption (I') (and hence is considerably stronger than the assumptions we employed in §2). Making the same transformation of variables with respect to his assumption of a constant rate of generic mutation, we find that $n_k = e^{p_g t}$. We can then compute $\alpha(k)$ (which will now vary with k) by taking the derivative of n_k with respect to k. We obtain

$$\alpha(k) = p_g e^{(p_g - p_s)t}/p_s. \tag{53}$$

If we substitute these values in eq. (40) of case II, where we assumed slowly changing α, we find in the limit, as $t \to \infty$,

$\rho = p_g/p_s$, as required. Hence, we see that the process of §2 is essentially the same as the one treated by Yule.

It is interesting and a little surprising that when Yule, some twenty years after this discovery, examined the statistics of vocabulary, he did not employ this model to account for the observed distributions of word frequencies. Indeed, in his fascinating book on *The Statistical Study of Literary Vocabulary* (1944) he nowhere refers to his earlier paper on biological distributions.

5. Conclusions

This paper discusses a number of related stochastic processes that lead to a class of highly skewed distributions (the Yule distribution) possessing characteristic properties that distinguish them from such well-known functions as the negative binomial and Fisher's logarithmic series. In §1, the distinctive properties of the Yule distribution were described. In §§2 and 3 several stochastic processes were examined from which this distribution can be derived. In §4, a number of empirical distributions that can be approximated closely by the Yule distribution were discussed, and mechanisms postulated to explain why they are determined by this particular kind of stochastic process. In the same section, the derivations of §§2 and 3 were compared with models previously proposed by Yule (1924) and Champernowne (1953) to account for the data on biological species and on incomes, respectively.

The probability assumptions we need for the derivations are relatively weak, and of the same order of generality as those commonly employed in deriving other distribution functions – the normal, Poisson, geometric and negative binomial. Hence, the frequency with which the Yule distribution occurs in nature – particularly in social phenomena – should occasion no great surprise. This does not imply that all occurrences of this empirical distribution are to be explained by the process discussed here. To the extent that other mechanisms can be shown also to lead to the same distribution, its common occurrence is the less surprising. Conversely, the mere fact that particular data conform to the Yule distribution and can be given a plausible interpretation in terms of the stochastic model proposed here tells little about the underlying phenomena beyond what is contained in assumptions (I) through (III).

CHAPTER 2

Some further notes on a class of skew distribution functions

This chapter was originally published in 1960 in reply to some objections that had been raised to the derivations and conclusions of ch. 1. It is included here, not for the purpose of pursuing the debate with Dr. Mandelbrot (anyone interested in the subsequent course of that debate may refer to *Information and Control*, vol. 4, September 1961, which contains two further rounds of interchange between Mandelbrot and Simon), but because it elaborates upon and generalizes some of the results of ch. 1.

In particular, §3 of ch. 2 defines a new stochastic process from which the Yule distribution, Fisher's log series distribution and the Poisson distribution can be derived by specializing the parameters. This generalization provides an additional method for studying the sensitivity of the steady-state distributions to perturbations in the underlying assumptions.

The first two sections of ch. 2 are concerned with the size of the parameter, ρ. We saw in ch. 1 that the value, $\rho = 1$ corresponds to the limiting case in which the rate at which new units are entering the population approaches zero (zero rate of entry of new firms). The initial model of ch. 1 does not permit values of ρ less than 1. Since observed values often lie very close to that value, and some fitted values below it, it is highly desirable to find conditions under which this limitation on the theoretical value of ρ can be relaxed. Several such conditions are described in the present chapter.

The new stochastic model of §3 is described, as are the

previous models, in terms of word frequencies. We should like here to discuss its interpretation as a model for firm-size distributions. The model most closely resembles the model of ch. 1, §3, in that both incorporate death processes as well as birth processes, and as a consequence, both possess a zero-growth steady state. (In terms of firm sizes, both assume that aggregate sales of the totality of firms are constant.) The differences between the two models lie in the structure of the "death" terms. In ch. 1 it was assumed that all units had an equal probability of total extinction. In the present chapter it is assumed that each unit has a probability of losing sales that is proportionate to its current size; hence that both gains and losses of sales are governed by a Gibrat assumption or some generalization of it.

The argument of §3 is simplified if we consider the following specialization of eq. (3),

$$
\begin{aligned}
f(i, m+1) - f(i, m) = {} & [(1-\alpha)/k][(i-1) \\
& \times f(i-1, m) - if(i, m)] \\
& - [1/k][if(i, m) \\
& - (i+1)f(i+1, m)].
\end{aligned} \tag{3'}
$$

The first term on the right-hand side of (3') describes the change in firm-size frequencies when sales are transfered to a firm; the second term the change when sales are transfered from a firm. Both terms satisfy the Gibrat assumption. For a static solution, we assume the left side of (3') equals zero, a sufficient condition for which is: $if(i) = (i-1)f(i-1)$. This condition immediately yields the solution,

$$
f(i) = [f(1)/(1-\alpha)](1-\alpha)^i / i \tag{4'}
$$

This is Fisher's log series distribution, so named because the expression for each successive value of i is identical with the corresponding term of the series expansion of

$$
\begin{aligned}
-\log \alpha &= -\log (1 - (1-\alpha)) \\
&= (1-\alpha) + (1-\alpha)^2/2 + (1-\alpha)^3/3 + \cdots,
\end{aligned}
$$

where, as before, $0 < \alpha < 1$ measures the rate at which new units are entering the system (and old units leaving).

As can be seen from eq. (4'), the expression for $f(i)$ is increasingly dominated by the exponential factor as i increases, so that the distribution approaches the geometric distribution asymptotically in the upper tail. When plotted on log–log paper, the log series distribution shows a strong concavity to the origin, and would be unlikely to give a reasonable fit to firm size data for a growing economy. It is a likely candidate, however, to fit the distribution in an economy whose total size is constant. (See Vining (1975) for an analysis of a no-growth model as applied to a city size distribution.) The more general process of eq. (3), §3, has two additional parameters, d and c, that would allow it to be fitted to distributions whose mechanisms do not quite satisfy the Gibrat assumption.

1. Introduction

In a recent note, Dr. Benoit Mandelbrot (1953) has raised some objections to a stochastic explanation of certain well-known data on word frequencies. A number of fundamental points in the note appear incorrect, others are debatable. Some of these relate to the empirical properties of the distributions, some to the mathematical analysis. Since the words frequency data have attracted a great deal of attention, it is perhaps worthwhile to try to clarify the points at issue.

Let $f(i, k)$ be the number of different words, each of which occurs exactly i times, in a sample of k words of text. In a wide range of cases, the observed data can be fitted quite well by a function of the form,

$$f(i, k) = C(k)i^{-(\rho+1)}, \tag{1}$$

and even more satisfactorily, particularly for low values of i, by the function,

$$f(i, k) = AB(i, \rho + 1), \tag{2}$$

where A and ρ are constants, and $B(i, \rho + 1)$ is the beta function of $i, \rho + 1$. As i increases, (2) approaches (1) asymptotically.[1] For both (1) and (2), the expected value of i is finite if and only if $\rho > 1$.

Function (1) has a long history in statistics; in economics, it is usually associated with the name of Pareto, in linguistics, with the names of Estoup and Zipf. Zipf was particularly interested in the case where $\rho = 1$. Function (2) was first introduced by Yule (1924) to explain certain taxonomic data of Willis, and hence we have proposed calling it the Yule Distribution.

2. The empirical distributions

A great deal of Dr. Mandelbrot's critical discussion depends on his claim that for the empirical word-frequency distributions, $\rho < 1$. He states categorically (1959, p. 92): "One finds, in general, that $\rho < 1$ for word frequencies . . . The few cases where $\rho > 1$ are also quite exceptional in other respects (e.g. Modern Hebrew about 1935)."

He makes an almost identical statement on page 498 of (1953). Unfortunately, he does not in either case present his evidence, and the source, Zipf, on which he chiefly relies, contradicts him. The data that Zipf report show ρ to be greater than 1 more often than not, and *almost always to be very close to 1* – a point to which we shall return. We find in Zipf the following least-squares estimates of ρ: for Joyce's *Ulysses*, between 0.99 and 1.01 (p. 34); Plautus, 0.98 (p. 34); the *Iliad*, 1.15 (p. 34); Nootka and Plains Cree holophrases, 1.36 and 1.14, respectively (p. 84); Nootka morphenes, 0.67 (p. 85); Nootka varimorphs, two values, 0.67 or 1.12, depending on the curve-fitting method (p. 85); Dakota words, 1.29 (p. 86); Gothic words, 1.025 (p. 94); old high German words, 0.98 (p. 116). In addition, a large number of values, all close to 1, are reported for children's speech.

In addition to the calculated values, Zipf presents a large number of graphs of distributions, on a double-log scale, in

---

[1]For additional detail see ch. 1, p. 27.

virtually all of which ρ is very close to 1 – sometimes a little greater, sometimes a little less. The figure on page 25 of Zipf, for example, strikingly conforms to the hypothesis, with ρ indistinguishable from unity. In most of the distributions (see, e.g., Zipf, pp. 123, 125), there is a little curvature, usually a convexity upward. Under these circumstances, neither function (1) nor (2) fits exactly, and it is difficult to know how best to estimate ρ. An unweighted least-squares fit to the distribution on a logarithmic scale is perhaps not the most plausible method.

Several estimators are proposed in ch. 1. If k is the size of sample, n_k the number of different words in the sample, and $f(1)$ the number of different words each of which occurs exactly once, then, by eqs. (25) and (18) of ch. 1, we have $\alpha = n_k/k$ and $\rho = 1/(1 - \alpha)$. Using these relations, we find the following values for ρ: *Ulysses*, 1.13; Eldridge's word count, 1.16; Yule's count of nouns in Macaulay, 1.33; Plautus, 1.34. (In fairness, it should be pointed out that when this method of estimating is used, ρ is necessarily greater than 1.) Alternatively, we can estimate α by eq. (27): $(2 - \alpha) = n_k/f(1, k)$. We then find the following values: *Ulysses*, 1.24; Eldridge, 0.983; Macaulay, 0.935; Plautus, 1.81. (See the discussion of this estimator on page 35 of ch. 1.)

Finally, it should be observed that if $\rho < 1$, neither (1) nor (2) can hold through the entire range, for in this case the mean of the distribution would be infinite. No model (and this applies to Dr. Mandelbrot's as well as to ours) that requires $\rho < 1$ can hold for indefinitely large values of i. Empirically this shows up in the curvature of the observed distributions for large i.

We must conclude that Dr. Mandelbrot has not established his case that, in general, $\rho < 1$. On the contrary, the data suggest that generally $\rho \sim 1$. But what is the significance of this? Several derivations of (1) in Mandelbrot (1953, 1954) require that $\rho < 1$ (page 495), and therefore fail to handle any of the empirical distributions for which the parameter exceeds unity. On the other hand, the *first* derivation (pp. 28–31) of (2) in ch. 1 requires that $\rho > 1$, and therefore fails when the parameter falls short of unity. However, a number of variant models are discussed in ch. 1 which lead, approximately, to (2), and which admit $\rho < 1$. We shall discuss below whether these variants involve "analytic circularity" (Dr. Mandelbrot's term for "lack of parsimony").

In trying to decide whether the parameter is greater than or less than unity, we must not lose sight of the striking fact, already mentioned, that it is almost always very close to unity. It is hard to specify *how* close for there are no satisfactory tests of closeness of fit in these matters, and hence it is not surprising that different statisticians, equally "skilled in the art," may experience different degrees of satisfaction with the results. I. J. Good (1953, pp. 258–259), for example, after fitting (2), in the special case where $\rho = 1$, to the Eldridge word-frequency count, concludes that the fit "is remarkably good" for $i \leqslant 15$, and can be improved by introducing a convergence factor. He fits the same function to Yule's sample of nouns in Macaulay's essay on Bacon, and says (p. 261): "It is curious that this should again give such a good fit for values of i that are not too large ($i \leqslant 30$). The sample is of nouns only and, moreover, Yule took different inflexions of the same word as the same."

Yule, himself, was much more critical, rejecting the fit of (1) to both Zipf's data and his own (1944, p. 55): "I spent some time on a re-examination of his data and cannot agree with the claim that the formula holds to any satisfactory degree of precision even for his distributions: it certainly does not hold for any of my own that I have tested."[2]

If we accept Mr. Good's more optimistic conclusion that some of the fits are "remarkably good" for the limiting case, where we take $\rho = 1$, then we would like our theory of the phenomena to explain the special significance of this limiting case. The derivation of (2) in ch. 1 does this, for it shows that as long as the ratio of number of *different* words in the text to total word occurrences is small (say, not more than 0.2), the parameter will be close to 1 (say, not over 1.25).

Before leaving the subject of the empirical distributions, we should like to state our agreement with Dr. Mandelbrot that for

[2]We would conjecture that Yule used the chi-square test to reject the hypothesis. We are confronted here with the usual difficulties of testing an extreme hypothesis. (See ch. 6, below.)

the taxonomic examples of Willis, $\rho < 1$, for income distributions, $\rho > 1$. But the data on pages 377–382 of Zipf clearly contradict his assertion that "for non-biological taxonomies such as names of professions, business catalogues, etc., ... ρ is always less than one, and usually it is close to $\frac{1}{2}$."

3. The stochastic models

In §2 of ch. 1, we formulated a stochastic model that yields (2) as its steady-state distribution.[3] As we pointed out there, the definition of "steady state" poses some difficulties. Hence, we reinterpreted the same model in §3, by means of an alternative urn scheme, in a way that allowed a rigorous definition of "steady state."

Dr. Mandelbrot's principal objections, however, are levelled against the derivations in the case where $\rho < 1$. We have already given the reasons from empirical observation for thinking this is not generally the significant case for word frequencies. Nevertheless, this case certainly does arise in some instances, (e.g. the Thorndike count), and in applications of these kinds of stochastic models to other data (e.g. the taxonomic data of Willis). Hence, we should like to discuss this case a little more fully.

On pages 34–36 of ch. 1, we show heuristically how the case $\rho < 1$ for small i might arise. Dr. Mandelbrot (1959, p. 96), after introducing several approximations, which he does not justify in

[3]Since Dr. Mandelbrot mentions several times that this model is a special case of Champernowne's, I should like to put the record straight. Champernowne never derives (2), but only the approximation, (1). Yule derives (2) for the case $\rho < 1$, but not for $\rho > 1$. Neither Champernowne's derivation nor Yule's discloses the special significance of the limiting case, $\rho = 1$, or the reasons why the word distributions should lie close to this limiting case. Moreover, the assumptions required for my derivation of (2) are much weaker than Yule's. Finally, since Rapoport (1957, p. 157) has suggested that my derivation was a "counter-analysis" to Mandelbrot's, I might mention that at the time I derived (2) I was not familiar with the papers of Mandelbrot, Champernowne, or Yule. I came across these in the course of the search for prior work that one normally makes before publishing. [H.A.S.]

detail, shows that our approximation can be "exact" only in a very special case. We will go further, and say (as we did already on page 33 of ch. 1 that it cannot be exact even in that special case because of nonconvergence as i increases.

On page 49 of ch. 1, we gave a short sketch of an alternative derivation of (2) for $\rho < 1$, corresponding to Yule's (1924) original case. Yule himself obtained convergence by using the incomplete beta function, for the upper limit of integration that then appears as an additional parameter can be given a natural interpretation in his application. It also has a natural interpretation for our urn models, for we must have $f(i) = 0$ for all $i > k$. When k is very large, as it is for the word frequency samples, there appears to be ample justification for ignoring this restriction to gain analytic simplicity.

Neither of the derivations mentioned in the last two paragraphs makes use of the more rigorous method of §3 of ch. 1, we should like to exhibit a stochastic process that has as its steady state a function that is a slight generalization of (2), and that has $\rho < 1$. The method of formulating this process will provide some additional evidence for our original contention that there is a whole host of stochastic processes that yield equilibrium distributions quite similar to the observed ones, and that, therefore, we should be wary about concluding much more than that some law of large numbers is at work.

We consider a sequence of k words. We add words to the end of the sequence, and drop words from the sequence at the same average rate, so that the length of the sequence remains k.[4] For the birth process we assume: the probability that the next word added is one that now occurs i times is proportional to $(i + c)f(i)$. The probability that the next word is a new word is α, a constant. For the death process we assume: the probability that the next word dropped is one that now occurs i times is proportional to $(i + d)f(i)$. The terms c and d are constant parameters. The steady-state relation is:

[4]For details, see pages 36–38 of ch. 1.

$$f(i, m+1) - f(i, m) = \frac{(1-\alpha)}{k + cn_k} [(i-1+c)$$
$$\times f(i-1, m) - (i+c)f(i, m)]$$
$$- \frac{1}{k + dn_k} [(i+d)f(i, m)$$
$$- (i+d+1)f(i+1, m)] = 0. \quad (3)$$

A solution to this equation, independent of m, is

$$f(i, m) = A\lambda^i B(i+c, d-c+1), \quad (4)$$

where
$$\lambda = \frac{(1-\alpha)(k + dn_k)}{(k + cn_k)},$$

and B is the beta function. If we compare (4) with (2), we see that the latter has a convergence factor, λ, that is missing from the former, and that ρ has been replaced by $d - c = \rho^*$. In particular, if d is not much larger than c, we will have $\rho^* < 1$.

The process (3) has a number of interesting special and limiting cases. For example, if $c = d$, the steady-state distribution is a generalization of Fisher's log series distribution: $f(i) = A(1-\alpha)^i/(i+c)$, where A is a normalizing constant. On the other hand, as d approaches zero and c increases without limit, we obtain the limiting process,

$$f(i, m) = \frac{(1-\alpha)}{n_k} [f(i-1) - f(i)]$$
$$- \frac{1}{k} [if(i) - (i+1)f(i+1)] = 0, \quad (5)$$

the steady-state distribution for which is simply the Poisson distribution: $f(i) = A\lambda^i/i!$ The reader can verify these results, by substituting the solutions in eqs. (3) and (5), respectively.

4. The meaning of the word frequency distributions

It appears from this analysis that the stochastic interpretation of the word frequency data proposed in ch. 1 is decidedly more adequate than Dr. Mandelbrot allows. What is the relation of this interpretation to the alternative interpretations that Dr. Mandelbrot had proposed (1953, 1954)? Dr. Mandelbrot's models are of two types:

(1) Derivations of the distribution from various assumptions of efficient letter-by-letter coding of the language.
(2) Derivations of the distribution from various Markovian assumptions about the stochastic formation of words from strings of letters.

From a formal mathematical standpoint, Dr. Mandelbrot's efficient coding models and his stochastic models are substantially equivalent. The two types of derivations correspond, respectively, to derivations in classical statistical mechanics based on entropy maximization, on the one hand, and statistical equilibrium, on the other. Dr. Mandelbrot's stochastic models are quite different from those of ch. 1, since the latter rest on no assumptions whatsoever about the statistical properties of the alphabet in which the words are encoded.

It seems to us something more than a matter of taste and convenience whether certain empirical regularities can be explained as the products of stochastic processes arising from imitation and association, as proposed in ch. 1; whether we explain them by postulating a mechanism that maximizes the amount of information transmitted per symbol; or whether we explain them on the basis of statistical properties of the encoding process. Our feeling that the teleological explanations are particularly to be avoided unless other evidence requires them is perhaps a prejudice, but it is a prejudice shared by others. Miller, Newman, and Friedman (1958) say, for example: "This derivation [the one numbered (2) above] has the advantage that it does not assume optimization in terms of cost; it begins with the more palatable assumption that the human source is a stochastic process."

As between the two stochastic explanations, we confess also a preference for that developed in ch. 1. First, unlike the stochastic derivation from coding considerations, it involves mechanisms of imitation and association that are consistent with what we know about social and psychological processes. Second, while all the data on the word frequency distribution show it to be extremely regular, the data on the variation of word frequency with word length show only a very rough relation. This suggests that very frequent words become abbreviated in use, and hence generally become short words. Use causes shortness, not shortness use. Common sense suggests the same thing. However, it would be nice to be able to choose between the two major types of stochastic models on the basis of clearcut evidence rather than these very crude considerations. The evidence remains to be discovered.

CHAPTER 3

Properties of the Yule distribution

This chapter has not previously been published. It derives a number of the most important properties of the Yule distribution, including the conditions under which it has a finite mean and a finite variance. It also derives the maximum likelihood estimate of the parameter.

The final pages of the chapter examine the departures of the Yule distribution from linearity, when the cumulative distribution function is plotted on log–log paper. It is shown that the distribution is slightly concave to the origin (convex upwards) on a log–log scale, throughout its range, for $\rho > 1$, but slightly convex to the origin (concave upwards) for $\rho < 1$. The nonlinearity is very slight, however, even for values of ρ as small as 0.1 and as large as 10.

1. The Yule distribution

A number of highly skewed size distributions, such as corporate assets, city populations, individual incomes, word frequencies in a text, have a density function proportional to a complete beta function

$$B(i, \rho + 1) = \int_0^1 \tau^{i-1}(1 - \tau)^\rho \, d\tau, \tag{1}$$

where i, the random variable for size, takes positive integer values and ρ is a constant. This skew distribution, whose density is

$$f(i) = AB(i, \rho + 1), \tag{2}$$

where A is a normalizing constant, was named the Yule distribution in ch. 1, and a process generating it was described.

This process starts with a few elements all of unit size in order to initialize the population. At each epoch, τ, the aggregate size of the population (the sum of the sizes of the elements of the population) is increased by one unit. Two assumptions govern the rule on which element is to receive the unit increment. The first assumption states that there is a constant probability that the unit goes to a new element not previously in the population. If this happens, a new element of unit size is created in the population. If the unit does not go to a new element, it goes to an old element. In this case, the selection of the old element to which the unit is given is governed by the second assumption: the probability that the selected element is of size i (before the unit is added) is proportional to the aggregate size of all elements with size i (i times the number of elements with size i). The second assumption incorporates Gibrat's law of proportionality. Hence, big or small, each element has the same chance of growing at any given percentage in a given period.

The purpose of this chapter is to explore some of the properties of the Yule distribution so that, when, it is fitted to empirical data, these properties can be used to understand the nature of empirical data more concretely.

2. The distribution function

We first sum the beta function over all integer values of i greater than or equal to i,

$$\sum_{j=i}^{\infty} B(j, \rho + 1) = \sum_{j=i}^{\infty} \int_0^1 \tau^{i-1}(1 - \tau)^\rho \, d\tau$$

$$= \int_0^1 (1 - \tau)^\rho \left[\sum_{j=i}^{\infty} \tau^{i-1} \right] d\tau$$

$$= \int_0^1 (1 - \tau)^\rho \frac{\tau^{i-1}}{1 - \tau} \, d\tau = B(i, \rho). \tag{3}$$

Then, let us define a distribution function, $F(i)$, cumulative from the right, namely

$$F(i) \equiv \sum_{j=i}^{\infty} f(i) = A \sum B(j, \rho + 1) = AB(i, \rho). \tag{4}$$

Now, since $i = 1$ is the minimum size of elements in the population, $F(1) \equiv 1$, hence

$$F(1) = A \int_0^1 (1 - \tau)^{\rho-1} \, d\tau = - \frac{A(1 - \tau)^\rho}{\rho} \Big|_0^1 = A/\rho \equiv 1, \tag{5}$$

which means that $A = \rho$. Thus, the cumulative distribution function of the Yule distribution is

$$F(i) = \rho B(i, \rho), \tag{6}$$

and the density function is

$$f(i) = \rho B(i, \rho + 1). \tag{7}$$

Thus, clearly $\rho > 0$ is necessary for the distribution.

In order to analyze the properties of $F(i)$, let us first calculate $F(i + 1)/F(i)$. To do this, we express the beta function as a product of gamma functions using,

$$B(1, \rho) = \frac{\Gamma(i)\Gamma(\rho)}{\Gamma(i + \rho)}, \tag{8}$$

where

$$\Gamma(x) = \int_0^{\infty} e^{-\tau} \tau^{x-1} \, d\tau \tag{9}$$

(See, for example, Artin (1964, pp. 11, 19).)

The gamma function has the property

$$\Gamma(x + 1) = x\Gamma(x) \tag{10}$$

(see Artin (1964, p. 12).) Therefore,

$$B(i + 1, \rho) = \frac{i}{i + \rho} \cdot \frac{\Gamma(i)\Gamma(\rho)}{\Gamma(i + \rho)} = \frac{i}{i + \rho} \cdot B(i, \rho). \tag{11}$$

From this we obtain

$$\frac{F(i+1)}{F(i)} = \frac{i}{i+\rho}.$$
(12)

Since by definition $F(1) = 1$, we can derive $F(i)$ successively as follows:

$$F(1) = 1,$$

$$F(2) = \frac{1}{1+\rho},$$

$$F(3) = \frac{1}{1+\rho} \cdot \frac{2}{2+\rho},$$

$$\vdots$$

$$F(i) = \prod_{j=1}^{i-1} \frac{j}{j+\rho}.$$
(13)

It is easy to see that $F(i)$ is a decreasing function of ρ for all $i > 1$.

The density $f(i)$ is then calculated using (6), (8) and (10),

$$f(i) = F(i) - F(i+1) = \rho B(i, \rho) - \rho B(i+1, \rho)$$

$$= \rho \left[\frac{\Gamma(i)\Gamma(\rho)}{\Gamma(i+\rho)} - \frac{\Gamma(i+1)\Gamma(\rho)}{\Gamma(i+\rho+1)} \right]$$

$$= \rho \frac{\Gamma(i)\Gamma(\rho)}{\Gamma(i+\rho)} \left(1 - \frac{i}{i+\rho} \right)$$

$$= \frac{\rho}{i+\rho} F(i).$$
(14)

Also from (12), we have

$$f(i) = \frac{\rho}{i} F(i+1).$$
(15)

Thus,

$$f(1) = \frac{\rho}{1+\rho},$$

$$f(2) = \frac{1}{1+\rho} \cdot \frac{\rho}{2+\rho},$$

$$f(3) = \frac{1}{1+\rho} \cdot \frac{2}{2+\rho} \cdot \frac{\rho}{3+\rho},$$

$$\vdots$$

$$f(i) = \prod_{j=1}^{i-1} \frac{j}{j+\rho} \cdot \frac{\rho}{i+\rho}. \tag{16}$$

The parameter, ρ, of the Yule distribution can easily be determined if $f(1)$ or $F(2)$ is known, since

$$\rho = \frac{f(1)}{1-f(1)} = \frac{1}{F(2)} - 1. \tag{17}$$

3. The mean and variance

The mean μ of a Yule distribution exists if and only if $\rho > 1$ and is given by

$$\mu = \sum_{i=1}^{\infty} if(i) = \sum (iF(i) - iF(i+1)) = \sum F(i)$$

$$= \rho \int_0^1 \sum \tau^{i-1}(1-\tau)^{\rho-1} \, d\tau = \rho \int_0^1 (1-\tau)^{\rho-2} \, d\tau$$

$$= -\frac{\rho(1-\tau)^{\rho-1}}{\rho-1} \bigg|_0^1 = \frac{\rho}{\rho-1}. \tag{18}$$

Note that (18) implies

$$\frac{1}{\mu} + \frac{1}{\rho} = 1. \tag{19}$$

The variance of i, $\text{var}(i)$, exists if and only if $\rho > 2$. To calculate it, first we derive, using $\sum_{i=1}^{\infty} i\tau^{i-1} = 1/(1-\tau)^2$,

$$\sum_{i=1}^{\infty} iF(i) = \rho \int_0^1 \sum i\tau^{i-1}(1-\tau)^{\rho-1} \, d\tau = \rho \int_0^1 (1-\tau)^{\rho-3} \, d\tau$$

$$= -\frac{\rho(1-\tau)^{\rho-2}}{\rho-2} \bigg|_0^1 = \frac{\rho}{\rho-2}. \tag{20}$$

Then, using (15), (18), and (20),

$$\mathrm{var}(i) = \sum i^2 f(i) - \mu^2$$

$$= \sum i\rho F(i+1) - \mu^2$$

$$= \rho \left[\sum iF(i) - \sum F(i) \right] - \mu^2$$

$$= \frac{\rho^2}{\rho - 2} - \frac{\rho^2}{\rho - 1} - \frac{\rho^2}{(\rho - 1)^2}$$

$$= \rho^2 \frac{\rho^2 - 2\rho + 1 - (\rho^2 - 3\rho + 2) - (\rho - 2)}{(\rho - 2)(\rho - 1)^2}$$

$$= \frac{\rho^2}{(\rho - 2)(\rho - 1)^2} = \frac{\mu^2}{\rho - 2}. \tag{21}$$

4. Maximum likelihood estimate of ρ

The maximum likelihood estimate of ρ is derived as follows. For a sample of n elements whose sizes are $i_1, i_2, \ldots, i_n$, we obtain the likelihood function

$$L = \prod_{k=1}^n f(i_k) = \prod_{k=1}^n \rho B(i_k, \rho + 1)$$

$$= \prod_{k=1}^n \rho \cdot \frac{\Gamma(i_k)\Gamma(\rho + 1)}{\Gamma(i_k + \rho + 1)}. \tag{22}$$

Since $\log L$ is a monotone increasing function of L, we calculate $\mathrm{d}\log L / \mathrm{d}\rho$ and set it equal to zero (all logarithms in this paper are natural logs unless the contrary is stated),

$$\frac{\mathrm{d}\log L}{\mathrm{d}\rho} = n\frac{\mathrm{d}\log \rho}{\mathrm{d}\rho} + n\frac{\mathrm{d}\log \Gamma(\rho + 1)}{\mathrm{d}\rho}$$

$$- \sum \frac{\mathrm{d}\log \Gamma(i_k + \rho + 1)}{\mathrm{d}\rho}. \tag{23}$$

Since from (10)

$$\frac{d \log \Gamma(x+1)}{dx} - \frac{d \log \Gamma(x)}{dx}$$

$$= \frac{d(\log x + \log \Gamma(x))}{dx} - \frac{d \log \Gamma(x)}{dx} = \frac{1}{x}, \quad (24)$$

we have by applying (24) repeatedly,

$$\frac{d \log \Gamma(x+k)}{dx} - \frac{d \log \Gamma(x)}{dx}$$

$$= \frac{1}{x} + \frac{1}{x+1} + \cdots + \frac{1}{x+k-1}. \quad (25)$$

Using (25)

$$\frac{d \log L}{d\rho} = \frac{n}{\rho} - \sum_{k=1}^{n} \left(\frac{d \log \Gamma(i_k + \rho + 1)}{d\rho} - \frac{d \log \Gamma(\rho + 1)}{d\rho} \right)$$

$$= \frac{n}{\rho} - \sum \left(\frac{1}{1+\rho} + \frac{1}{2+\rho} + \cdots + \frac{1}{i_k + \rho} \right). \quad (26)$$

If we let $f^a(i)$ be the actual proportion of elements with size i in the sample, $nf^a(i)$ be the actual frequency of elements in the sample, and $F^a(i)$ be the actual proportion of elements with size i or greater in the sample, we can rewrite (26) as

$$\frac{d \log L}{d\rho} = \frac{n}{\rho} - \left(\frac{nF^a(1)}{1+\rho} + \frac{nF^a(2)}{2+\rho} + \cdots \right). \quad (27)$$

By setting this equal to zero in order to maximize L, we have

$$\sum_{i=1}^{\infty} F^a(i) \frac{\rho}{i+\rho} = 1. \quad (28)$$

There is a unique solution ρ^* to (28) for any sequence $\{F^a(i)\}$, where $1 = F^a(1) \geqslant F^a(2) \geqslant F^a(3) \cdots \geqslant 0$, except for a degenerate sequence $F^a(1) = 1$, $F^a(i) = 0$, for $i = 2, 3 \cdots$ (all elements in the sample having a unit size), which yields $\rho = \infty$ to maximize L. Let us, therefore, assume that $F^a(2) > 0$. Now it is easy to see from (27) that $(d \log L)/d\rho \to +\infty$ as $\rho \to 0$. On the other hand for any $F^a(2) > 0$, there is a constant $\hat{\rho}$ such that for any $\rho > \hat{\rho}$, $(\mathrm{d} \log L)/\mathrm{d}\rho < 0$. To see this

$$
\begin{aligned}
\frac{\mathrm{d} \log L}{\mathrm{d}\rho} &= n \left(\frac{1}{\rho} - \sum \frac{F^a(i)}{i + \rho} \right) \leq n \left(\frac{1}{\rho} - \frac{1}{1 + \rho} - \frac{F^a(2)}{2 + \rho} \right) \\
&= n \frac{2 + \rho - \rho F^a(2) - \rho^2 F^a(2)}{\rho(1 + \rho)(2 + \rho)} \\
&= \frac{n}{(1 + \rho)(2 + \rho)} \left[\frac{2}{\rho} + (1 - F^a(2)) - \rho F^a(2) \right] \\
&< 0, \qquad \text{for } \rho > [2 + (1 - F^a(2))]/F^a(2).
\end{aligned}
\tag{29}
$$

Since $\mathrm{d} \log L /\mathrm{d}\rho$ is continuous on ρ, this shows that there is at least one ρ which satisfies (28). In addition, at $\rho = \rho^*$,

$$
\begin{aligned}
\left. \frac{\mathrm{d}^2 \log L}{\mathrm{d}\rho^2} \right|_{\rho=\rho^*} &= -\frac{n}{\rho^{*2}} + \sum \frac{n F^a(i)}{(i + \rho^*)^2} \\
&= \frac{n}{\rho^{*2}} \left(\sum F^a(i) \left(\frac{\rho^*}{i + \rho^*} \right)^2 - 1 \right) \\
&< \frac{n}{\rho^{*2}} \left(\sum F^a(i) \frac{\rho^*}{i + \rho^*} - 1 \right) = 0
\end{aligned}
\tag{30}
$$

by (28). This proves that ρ^* is not only the point at which L is maximized but also is the unique point with this property.

Also, the condition (28) for ρ^* can be rewritten using (14) as

$$
\sum_{i=1}^{\infty} f(i) \frac{F^a(i)}{F(i)} = 1,
\tag{31}
$$

meaning ρ^* makes the expected value of the ratio of actual and theoretical cumulative frequencies equal to 1.

5. The slope parameter

Many of the empirical skew distributions have been analyzed graphically by plotting the size of an element against its rank in the population (the rank of the largest element is set equal to 1) on a log–log scale. Therefore, it is interesting to see how the

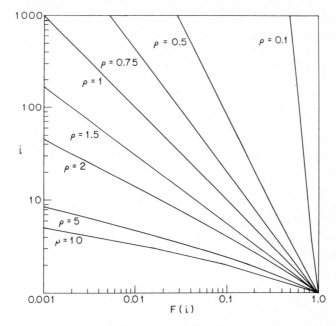

Fig. 3.1. Slope of Yule distribution as a function of the parameter ρ.

Yule distribution behaves when log $F(i)$ is plotted against log i (see fig. 3.1).

To determine the slope β of the distribution function on the log–log scale betweeen i and $i + 1$, we calculate

$$\beta(i) = -\frac{\log F(i + 1) - \log F(i)}{\log (i + 1) - \log i} = -\frac{\log (F(i + 1)/F(i))}{\log [(i + 1)/i]}$$

$$= -\frac{\log [(i/(i + \rho)]}{\log [(i + 1)/i]} = \frac{\log (1 + \rho/i)}{\log (1 + 1/i)}. \tag{32}$$

Note that $\beta(i) \geqslant 0$ for all $i = 1, 2, \ldots$. Thus, β may be expressed as a function of ρ and i implicitly, as in

$$\rho = \frac{(1 + 1/i)^\beta - 1}{1/i}, \tag{33}$$

which is the same formula as the future value of an annuity lasting for β periods at the interest rate $1/i$.

From (32) it may be seen that

$$\lim_{i \to \infty} \beta(i) = \rho, \tag{34}$$

by applying the l'Hospital's rule. Also, from (32) we have

$$\beta(1) = \frac{\log(1 + \rho)}{\log 2} = \log_2(1 + \rho). \tag{35}$$

The parameter ρ can, therefore, be determined by

$$\rho = 2^{\beta(1)} - 1 \tag{36}$$

if $\beta(1)$ is known.

Clearly, $\beta = 1$, if $\rho = 1$. We shall show that β is an increasing function of i if $\rho > 1$ and a decreasing function of i if $\rho < 1$. Let us first define

$$\phi = \log(1 + \rho/i),$$
$$\psi = \log(1 + 1/i). \tag{37}$$

Then, using $\rho/i = e^{\phi} - 1$ and $1/i = e^{\psi} - 1$,

$$\frac{d\phi}{di} = \frac{-\rho/i^2}{1 + \rho/i} = -\frac{(e^{\phi} - 1)(e^{\psi} - 1)}{e^{\phi}},$$
$$\frac{d\psi}{di} = \frac{-1/i^2}{1 + 1/i} = -\frac{(e^{\psi} - 1)^2}{e^{\psi}}. \tag{38}$$

Based on the definition of β in (32), $\beta = \phi/\psi$, hence

$$\frac{d\beta}{di} = \frac{\partial\beta}{\partial\phi} \cdot \frac{d\phi}{di} + \frac{\partial\beta}{\partial\psi} \cdot \frac{d\psi}{di} = -\frac{(e^{\phi} - 1)(e^{\psi} - 1)}{\psi \, e^{\phi}} - \frac{\phi(e^{\psi} - 1)^2}{\psi^2 \, e^{\psi}}$$
$$= \frac{\phi}{\psi}(e^{\psi} - 1)\left[\frac{1 - e^{-\psi}}{\psi} - \frac{1 - e^{-\phi}}{\phi}\right]. \tag{39}$$

We note that generally,

$$\frac{d}{dx}\left[\frac{1 - e^{-x}}{x}\right] = \frac{x \, e^{-x} - 1 + e^{-x}}{x^2} = \frac{e^{-x}(1 + x) - 1}{x^2} < 0, \tag{40}$$

because

$$e^x > 1 + x \quad \text{and} \quad e^{-x} < \frac{1}{1 + x}. \tag{41}$$

Therefore, if $\rho > 1$, $\phi > \psi > 0$, which means that $d\beta/di > 0$; while if $\rho < 1$, $\psi > \phi > 0$, which means that $d\beta/di < 0$.

This means that the slope β starts at $\log_2(1 + \rho)$ and increases if $\rho > 1$ or decreases if $\rho < 1$ as i is increased, approaching ρ as i goes to infinity. Therefore, the cumulative distribution is concave toward the origin if $\rho > 1$ and convex if $\rho < 1$. (See fig. 3.1.)

6. The Pareto distribution

The discrete Pareto distribution has the right cumulative distribution given by

$$F(i) = \sum_{j=i}^{\infty} f(i) = i^{-\rho}. \tag{42}$$

The Pareto distribution and the Yule distribution coincide when the parameter of distribution ρ is 1, since $F(i)$ in (13) is reduced to $1/i$ if $\rho = 1$. On the log–log scale,

$$\log F(i) = -\rho \log i, \tag{43}$$

namely the cumulative distribution is a straight line on the log–log scale. Furthermore,

$$\beta(i) = -\frac{\log F(i+1) - \log F(i)}{\log(i+1) - \log i} = -\frac{\log(F(i+1)/F(i))}{\log[(i+1)/i]}$$

$$= \rho \tag{44}$$

Thus, the parameter ρ of the distribution shows the slope of the distribution β as defined in (32) for Yule distribution.

To compare the two distributions, let $F_y(i)$ be the $F(i)$ defined in (13) for the Yule distribution and $F_p(i)$ be the $F(i)$ defined in (42) for the Pareto distribution with the same parameter ρ. Then, from (12) and (42),

$$\frac{F_y(i+1)}{F_y(i)} = \frac{i}{i+\rho} = \frac{1}{1+\rho/i}, \tag{45}$$

$$\frac{F_p(i+1)}{F_p(i)} = \left(\frac{i+1}{i}\right)^{-\rho} = \frac{1}{(1+1/i)^\rho}. \tag{46}$$

Thus, if $\rho > 1$, $1 + \rho/i < (1 + 1/i)^\rho$, hence $F_y(i) > F_p(i)$ for all i except $i = 1$ at which $F_y(1) = F_p(1) = 1$; if $\rho < 1$, $1 + \rho/i > (1 + 1/i)^\rho$, so that $F_p(i) \geqslant F_y(i)$.

Furthermore, by the property of the gamma function,

$$\lim_{i \to \infty} \Gamma(i)/\Gamma(i + \rho) = i^{-\rho} \tag{47}$$

(see (Artin (1964)). Therefore,

$$\lim_{i \to \infty} [\log F_y(i) - \log F_p(i)] = \lim_{i \to \infty} \left[\log \rho \frac{\Gamma(i)\Gamma(1)}{\Gamma(i + \rho)} - \log i^{-\rho} \right]$$

$$= \log \rho \Gamma(\rho) = \log \Gamma(\rho + 1). \tag{48}$$

Thus, the difference between the two distributions on the log–log scale converges to a constant, $\log \Gamma(\rho + 1)$.

7. The incomplete beta function

Let us next explore the effect on the Yule distribution of changing the complete beta function in (6) to an incomplete beta function. Let

$$B_\theta(i, \rho) = \int_0^\theta \tau^{i-1}(1 - \tau)^{\rho-1} \, d\tau, \tag{49}$$

where $0 < \theta < 1$. The derivation in (3) still holds for $B_\theta(i, \rho)$, hence

$$F_\theta(i) \equiv \sum_{j=i}^{\infty} f_\theta(i) = A \sum B_\theta(j, \rho + 1)$$

$$= A B_\theta(i, \rho), \tag{50}$$

where the subscript θ is used to indicate the parameter in the incomplete beta function. Since we want to have $F_\theta(1) \equiv 1$,

$$F_\theta(1) = A \int_0^\theta (1 - \tau)^{\rho-1} \, d\tau = - \frac{A(1 - \tau)^\rho}{\rho} \Big|_0^\theta$$

$$= A(1 - (1 - \theta)^\rho)/\rho \equiv 1, \tag{51}$$

which means that $A = \rho/(1 - (1 - \theta)^\rho)$. Thus,

$$F_\theta(i) = \rho B_\theta(i, \rho)/(1 - (1 - \theta)^\rho). \qquad (52)$$

Next, let us consider the incomplete beta function ratio $l_\theta(i, \rho)$ given by

$$l_\theta(i, \rho) = \frac{B_\theta(i, \rho)}{B(i, \rho)} = \frac{1}{B(i, \rho)} \int_0^\theta \tau^{i-1}(1 - \tau)^{\rho-1} \, \mathrm{d}\tau. \qquad (53)$$

Using the relation $B_\theta(i, \rho) = B(i, \rho)l_\theta(i, \rho)$,

$$F_\theta(i) = \rho B(i, \rho) \frac{l_\theta(i, \rho)}{1 - (1 - \theta)^\rho} = F(i) \frac{l_\theta(i, \rho)}{1 - (1 - \theta)^\rho}. \qquad (54)$$

We then relate $l_\theta(i, \rho)$ with the negative binomial distribution. Let $\Pi(i; \rho, \theta)$ be the probability of having exactly i successes before ρ failures occur, when the probability of success is θ. Then,

$$\Pi(i; \rho, \theta) = \binom{i + \rho - 1}{i} \theta^i (1 - \theta)^\rho. \qquad (55)$$

To make the formula applicable even when ρ is not an integer, let us modify (55), using (10), to

$$\Pi(i; \rho, \theta) = \frac{\Gamma(i + \rho)}{i\Gamma(i)\Gamma(\rho)} \theta^i (1 - \theta)^\rho = \frac{\theta^i (1 - \theta)^\rho}{iB(i, \rho)}. \qquad (56)$$

Then, in general, by the properties of the incomplete beta function,

$$l_{(1-\theta)}(\rho, i) = \sum_{x=0}^{i-1} \Pi(x; \rho, \theta). \qquad (57)$$

Alternatively,

$$l_\theta(i, \rho) = 1 - l_{(1-\theta)}(i, \rho) = \sum_{x=i}^\infty \Pi(x; \rho, \theta). \qquad (58)$$

Since (58) is defined for noninteger values of ρ, we may consider (58) to be a generalization of the cumulative negative binomial distribution.

Now, by definition of a negative binomial distribution, we have $l_\theta(0, \rho) = 1$ and $l_\theta(0, \rho) - l_\theta(1, \rho) - \Pi(0; \rho, \theta) - (1 - \theta)^\rho$, for the probability of having no success before ρ failures occur. Let $\Pi_\theta(i)$ be the probability of having i successes or more before ρ failures occur, given that at least one success has occurred. Then

$$\Pi_\theta(i) = l_\theta(i, \rho)/(1 - (1 - \theta)^\rho). \tag{59}$$

Thus, from (54) and (59) we have

$$F_\theta(i) = F(i)\Pi_\theta(i). \tag{60}$$

From this we conclude that the Yule distribution based on an incomplete beta function $B_\theta(i, \rho)$ is a joint product of two processes, one described in §1 involving Gibrat's law of proportionality and the constant entry rate and the other described above, the negative binomial process with the probability of success equal to θ. Thus, $F_\theta(i)$ is the probability that an element in the population grows to size i or more based on the process described in §1 *and* gains the necessary growth before ρ failures occur when the probability of success (increasing the size by 1) is θ.

For $\rho = 1$, the negative binomial distribution becomes a geometric distribution. Therefore, $F_\theta(i)$ for $\rho = 1$ is a distribution obtained as the joint product of the Pareto distribution and the negative binomial distribution.

8. Random increments

Another variation of the Yule distribution involves a use of a geometric distribution in the underlying stochastic process. In the standard model discussed in ch. 1, the total size of the population is increased by one unit at each round, the unit being given either to an existing entity or to a new entity. Let us consider the effect on the Yule distribution of changing the increment to a discrete random variable J with a geometric distribution $\eta^{J-1}(1 - \eta)$, where η, $0 \leq \eta < 1$, is a parameter.

As in the standard model, the probability that this random increment will go to an existing entity of size i is proportional to i times the number of entities of size i, namely the aggregate size of the size i entities. Thus, the Gibrat's law of proportionality is maintained. Also, as in the standard model, there is a constant probability α that a new entity is created with size J.

The expected value of the increment is

$$\sum_{J=1}^{\infty} J\eta^{J-1}(1-\eta) = \left[\sum_{J=1}^{\infty} J\eta^{J-1}(1-\eta) \right.$$
$$\left. - \sum_{J=1}^{\infty} J\eta^{J}(1-\eta) \right] \Big/ (1-\eta)$$
$$= \sum_{J=1}^{\infty} \eta^{J-1} = 1/(1-\eta). \tag{61}$$

Also the probability that the increment will be J units or greater is

$$\sum_{j=J}^{\infty} \eta^{j-1}(1-\eta) = \eta^{J-1}. \tag{62}$$

We let $f(i, k)$ be the expected number of entities with size i after k rounds have been completed. We also let $F(i, k) = \sum_{j=i}^{\infty} f(j, k)$, namely the number of entities with size i or greater after k rounds.

At the $(k + 1)$st round, $F(i, k + 1)$ is greater than $F(i, k)$ for one of two reasons. One is that the $(k + 1)$st increment goes to a new entity (probability α) and the increment J turns out to be i or greater (probability η^{i-1}). The other is that the $(k + 1)$st increment goes to an old firm (probability $1 - \alpha$), an old firm of size j $(<i)$ is selected (probability $jf(j, k)/(k/(1 - \eta)) = (1 - \eta)j[F(j, k) - F(j + 1, k)]/k$, for the expected value of the aggregated size of all old firms is the expected value of increment $1/(1 - \eta)$ times the number of rounds which is k), and the increment turns out to be $i - j$ units or greater (probability η^{i-j-1}), where each of the three events are independent. Thus, we may write

$$F(i, k + 1) - F(i, k) = \alpha\eta^{i-1} + (1 - \alpha)(1 - \eta)\sum_{j=1}^{i-1}$$
$$\times j[F(j, k) - F(j + 1, k)]\eta^{i-j-1}/k. \tag{63}$$

However, at equilibrium, we expect $F(i, k)$ to grow in proportion to k. Hence,

$$F(i, k + 1)/F(i, k) = (k + 1)/k, \tag{64}$$

and the left-hand side of (63) becomes simply $F(i, k)/k$ when (64) is substituted in it.

Since the expected number of entities after k rounds is αk, the equilibrium cumulative distribution $F(i)$ is given by $F(i, k)/\alpha k$. Thus, we have from (63),

$$F(i) = \eta^{i-1} + (1 - \alpha)(1 - \eta) \sum_{j=1}^{i-1} j[F(j) - F(j + 1)]\eta^{i-j-1}. \tag{65}$$

However, from (65), by replacing $i + 1$ for i, we obtain

$$F(i + 1) = \eta^{i} + (1 - \alpha)(1 - \eta) \sum_{j=1}^{i} j[F(j) - F(j + 1)]\eta^{i-j}, \tag{66}$$

which is equivalent to the right-hand side of (65) times η, except that the second term in the right-hand side of (66) has an extra term $(j = i)$. Thus, we have:

$$F(i + 1) = \eta F(i) + (1 - \alpha)(1 - \eta)i[F(i) - F(i + 1)], \tag{67}$$

$$[(1 - \alpha)(1 - \eta)i + 1]F(i + 1)$$
$$= [(1 - \alpha)(1 - \eta)i + \eta]F(i), \tag{68}$$

hence,

$$F(i + 1)/F(i) = [(1 - \alpha)(1 - \eta)i + \eta]/$$
$$[(1 - \alpha)(1 - \eta)i + 1]. \tag{69}$$

Now we have $\rho = 1/(1 - \alpha)$ as defined in ch. 1, and we also replace η by $\zeta = \eta/(1 - \eta)$ for simplicity. Since $1/(1 - \eta) = 1 + \zeta$, ζ is the mean of the increment minus one, namely the extra units of increment in the expected value over and above the unit increment in the standard model. Then

$$F(i + 1)/F(i) = (i + \rho\zeta)/(i + \rho + \rho\zeta). \tag{70}$$

Using $F(1) = 1$,

$$F(i) = \sum_{j=1}^{i-1} \frac{j + \rho\zeta}{j + \rho + \rho\zeta}. \tag{71}$$

To express this in terms of the gamma functions, we use the relationship $\Gamma(x + 1) = x\Gamma(x)$, namely,

$$F(i) = \frac{\Gamma(i + \rho\zeta)\Gamma(1 + \rho + \rho\zeta)}{\Gamma(1 + \rho\zeta)\Gamma(i + \rho + \rho\zeta)} = \frac{\Gamma(i + \rho\zeta)\Gamma(\rho)}{\Gamma(i + \rho + \rho\zeta)}$$

$$\times \frac{\Gamma(1 + \rho + \rho\zeta)}{\Gamma(1 + \rho\zeta)\Gamma(\rho)} = \frac{B(i + \rho\zeta, \rho)}{B(1 + \rho\zeta, \rho)}. \tag{72}$$

Note that if $\eta = 0$, $\zeta = 0$, and the denominator of $F(i)$ becomes $B(1, \rho) = \Gamma(1)\Gamma(\rho)/\Gamma(1 + \rho) = 1/\rho$, hence $F(i) = \rho B(i, \rho)$ which is the standard case discussed in §2 above.

In order to understand the shape of $F(i)$, let us compare this $F(i)$ with a standard $F(i)$ given in (6) in which $\eta = 0$. We shall use $F_0(i)$ to denote the standard distribution where $\eta = 0$ and $F_\eta(i)$ for a distribution with $\eta > 0$. Since $(j + \rho\zeta)/(j + \rho + \rho\zeta) < j/(j + \rho)$ for all positive j, ρ, and ζ, we have $F_\eta(i) < F_0(i)$ for all $i > 1$, meaning that in a chart with $\log F(i)$ on the horizontal axis and $\log i$ on the vertical axis, $F_\eta(i)$ lies entirely above $F_0(i)$.

We may obtain the distribution $F_\eta(i)$ by replacing i in the standard model by $i - \rho\zeta$ and adjusting the normalizing factor, namely shifting the resulting chart to the right by an appropriate amount to make $F(1) = 1$.

This process may also be compared with another modification of the standard model where the unit increment is replaced by a constant increment of $(1 + \zeta)$ units. Under this modification, the resulting distribution, denoted by $F_\zeta(i)$, is obtained by replacing i in the standard model by $i(1 + \zeta)$. In terms of the chart, the new chart is obtained by simply shifting the standard chart upwards by $\log(1 + \zeta)$.

CHAPTER 4

Some distributions associated with Bose–Einstein statistics*

This chapter shows some further relationships between two common distribution functions: the geometric and the Yule. In physics, Bose–Einstein statistics are commonly used as the basis for the derivation of the geometric distribution. Hill (1974) showed that, under some rather complicated conditions, Bose–Einstein statistics can also yield the Pareto distribution, a limiting case of the Yule distribution. In this chapter, we describe a simple urn scheme exemplifying Bose–Einstein statistics, and show that with a minor change in the process either the Yule distribution or the geometric distribution can be derived from it. In the course of the discussion it is shown that the urn scheme for Bose–Einstein statistics satisfies the Gibrat assumption.

The difference between the two variants of the urn scheme, leading to the two different distributions, may be characterized as follows: if existing units may be split as well as new units created, the geometric distribution results. If new units may be created, but existing units may not be split, the Yule distribution results.

*We are grateful to Daniel R. Vining, Jr. for valuable comments on an earlier draft of this chapter. This work was supported by a research grant from the National Science Foundation.

1. Introduction

It is well known (Feller (1968, p. 61)) that under appropriate conditions the geometric distribution is a limiting distribution for Bose–Einstein statistics. Indeed, in applications to physics, it is usually treated as though it were the only limiting distribution. However, in a recent paper Hill (1974) has shown that under different (and much more complicated) conditions, Bose–Einstein statistics can be made to yield the Pareto distribution as a limiting distribution. Since the Pareto distribution often gives an excellent fit to data on the relative frequencies of cities of different sizes, Hill has offered his derivation of this distribution from Bose–Einstein statistics as an "explanation" of the observed city size distributions.

We do not intend to examine formally here the meaning of "explanation." Nevertheless, to say that cities are distributed by size according to the Pareto law *because* they obey Bose–Einstein statistics would appear, in common-sense terms, to be less an explanation than a relocation of the mystery. Cities change size as a result of the births and deaths that take place within them, of the migrations into them from nonurban or foreign places, and the migrations between pairs of them. A satisfying explanation of the observed size distributions in terms of Bose–Einstein statistics would have to show a relation between these statistics and the birth, death, and migration processes listed above.

There already exist in the literature several derivations of the Pareto law from plausible stochastic assumptions about birth, death, and migration processes (see chs. 1 and 2). Although they differ in their details, all of these derivations have in common some variant of what is often called Gibrat's law. This law, applied to cities, states that the accretions to the population of a city (or a group of cities of nearly equal population) by births and migration will occur at a rate (per capita) nearly independent of the present city size; and that the loss of inhabitants by deaths and migration will also occur at a rate nearly independent of city size. More generally, Gibrat's law postulates expected growth proportional to size. Under this assumption, we have something like a random walk on a logarithmic scale, and would expect to derive from it highly skewed limiting distributions related to the log normal. Indeed, this is what we find, for by small changes in the boundary conditions and other parameters of our stochastic process we can obtain the Yule distribution and its limiting form, the Pareto, as well as the log normal distribution, Fisher's logarithmic series, and the negative binomial.

In this paper, we will show first that Bose–Einstein statistics satisfy Gibrat's law. By this means, we will explain why the statistics of city size can be derived from Bose–Einstein statistics. Second, we will interpret the Bose–Einstein scheme in terms of stochastic processes to show what boundary conditions would lead to the geometric distribution, and what boundary conditions would lead to the Pareto distribution. Third, we will discuss the interpretations of the alternative assumptions in terms of city growth processes. Derivations of skewed distributions from Bose–Einstein statistics had apparently not been known until Hill's article appeared. Our derivation of the Pareto distribution from Bose–Einstein statistics will use much simpler methods, and much weaker assumptions, than those employed by Hill.

2. Bose–Einstein statistics and Gibrat's law

We use the familiar model of placing h objects called "stars" in n cells arranged in linear order as in $|***|*||**|$, two adjacent bars defining a cell. Let h_j be the number of stars in the jth cell from the left, and let $H = (h_1, h_2, \ldots, h_n)$ be a vector indicating a particular assignment of h stars in n cells, where $h = \sum_{j=1}^{n} h_j$. (In the above example, $H = (3, 1, 0, 2)$.) Bose–Einstein statistics assume that each star is indistinguishable from each other star, hence two such arrangements $H = (h_1, h_2, \ldots, h_n)$ and $H' = (h'_1, h'_2, \ldots, h'_n)$ with $\sum_{j=1}^{n} h_j = \sum_{j=1}^{n} h'_j = h$ are said to be indistinguishable if and only if $h_j = h'_j$ for all $j = 1, 2, \ldots, n$, and said to be distinguishable otherwise.

Then, Bose–Einstein statistics postulate that each distinguishable arrangement of h stars in n cells has an equal probability of occurrence. (For example, 2-0, 1-1, 0-2, each has probability $\frac{1}{3}$ instead of $\frac{1}{4}, \frac{1}{2}, \frac{1}{4}$ as under familiar Maxwell–Boltzman statistics.)

There are $(n + h - 1)!/(n - 1)!h!$ distinguishable arrangements of h stars in n cells, since this is the number of ways of arranging the $(n - 1)$ partitioning bars among h stars, with two additional fixed bars at the end of the array to define the end cells (Feller (1968, p. 38ff)). Hence, the probability of obtaining a given H is $(n - 1)!h!/(n + h - 1)!$.

We shall first show that Bose–Einstein statistics can be obtained from Gibrat's law of proportionality applied to the process of throwing in stars when the total number of cells is fixed at n. Let the size of the jth cell, k_j, be the number of stars in this cell plus one. Thus, the size may be interpreted as the number of spaces in the cell between two stars or between two bars or between a star and a bar. Then, the aggregate size of n cells is

$$k = \sum_{j=1}^{n} k_j = n + h. \tag{1}$$

We require under Gibrat's law of proportionality that the probability that the $(h + 1)$st star will fall in the jth cell be equal to k_j/k. Let

$$K = (k_1, k_2, \ldots, k_n) \tag{2}$$

be a vector of sizes of n cells after h stars have been thrown in. Let

$$K_j = (k_1, k_2, \ldots, k_j - 1, \ldots, k_n), \tag{3}$$

namely K with the jth component of K reduced by one. K can be obtained from K_j by getting the hth star in the jth cell, providing $k_j - 1 > 0$ since, by definition, the size of a cell can never be less than 1. Assume that Bose–Einstein statistics hold for arrangements of $h - 1$ stars in n cells. Then, the probability of obtaining K_j, denoted by $P(K_j)$, is

$$P(K_j) = (n - 1)!(h - 1)!/(n + h - 2)!, \quad \text{if } k_j - 1 > 0,$$
$$= 0, \quad \text{if } k_j - 1 = 0.$$
$$\tag{4}$$

The conditional probability of obtaining K given K_j is $(k_j - 1)/(n + h - 1)$. Therefore, the probability of obtaining K, denoted by $P(K)$, is

$$P(K) = \sum_{j=1}^{n} \frac{k_j - 1}{n + h - 1} P(K_j). \tag{5}$$

However, if $P(K_j) = 0$, then $k_j - 1 = 0$. Hence, $P(K_j)$ in (5) may be replaced by $(n - 1)!(h - 1)!(n + h - 2)!$ without affecting the value of $P(K)$, i.e.

$$P(K) = \frac{(n - 1)!(h - 1)!}{(n + h - 2)!} \cdot \sum_{j=1}^{n} \frac{k_j - 1}{n + h - 1} = \frac{(n - 1)!h!}{(n + h - 1)!}, \tag{6}$$

since $\sum_{j=1}^{n} k_j - 1 = n + h - n = h$. This shows that K is Bose–Einstein if the distribution before throwing in the hth star is Bose–Einstein and the hth star is placed according to Gibrat's law of proportionality. Since for $h = 0$, each cell is of size 1 and the Bose–Einstein condition is trivially satisfied, the distribution of h stars in n cells is always Bose–Einstein if stars are thrown in according to Gibrat's law.

Let us now consider a distribution, $f(i, n + h)$, of the number of cells with size i after h stars have been thrown in. Note that $\sum_{i=1}^{h+1} f(i, n + h) = n$. Then, there are $n!/\prod_{i=1}^{h+1}(f(i, n + h)!)$ arrangements that lead to the given size distribution and each of them has the same probability $(n - 1)!h!/(n + h - 1)!$ of occurrence under Bose–Einstein statistics. Hence, the probability of obtaining a given size distribution, denoted by $P(f)$, is

$$P(f) = \frac{n!}{\prod_{i=1}^{h+1}(f(i, n + h)!)} \frac{(n - 1)!h!}{(n + h - 1)!}. \tag{7}$$

Clearly, the probability is maximum when $f(i, n + h) = 0$ or 1 for all i, i.e. when no two cells are of equal size. Also, it is obvious that there is no steady state distribution. However, steady-state distributions can be obtained if the number of cells n is also allowed to increase proportionately as the number of stars is increased. Depending upon how new cells are created, we derive two distinct distributions which we shall discuss next.

3. Two limiting distributions for Bose–Einstein statistics

Let $f^*(i, k)$ be the probability that a cell will have size i when the aggregate size of all cells is k. Also let $f(i)$ be the steady-state probability that a cell will have size i, i.e. $f(i) = \lim_{k \to \infty} f^*(i, k)$.

Under certain boundary conditions, Gibrat's law is known to produce as its limiting distribution the Pareto distribution, given by

$$f(i) = Ci^{-(\rho+1)},\tag{8}$$

in which C, ρ are constant parameters. Under other boundary conditions, the limiting distribution for Bose–Einstein statistics is the geometric distribution

$$f(i) = (\rho - 1)\rho^{-i},\tag{9}$$

in which ρ is a constant parameter ($\rho > 1$). Eqs. (8) and (9) can be derived by considering the necessary conditions for a steady state of stochastic process based on Bose–Einstein statistics.

Consider, first, a process in which not only stars but also bars are added. At each round, either a bar or a star is selected with probability α and $1 - \alpha$, respectively. If a star is selected, it is thrown in according to Bose–Einstein statistics, so that each space has an equal chance of receiving it. If a bar is selected, however, it is placed next to an existing bar. That is to say, new cells are added at a rate α and all new cells are of unit size. The average size of cells is a random variable with mean $1/\alpha$.

Regardless of whether a bar or a star is selected at any given round, the aggregate size k of all cells is increased by one at the end of the round either because the size of one of the cells is increased by one or because a new cell of size 1 is added. Thus, we may use k not only as the aggregate size but also as a counter for the number of rounds.

Let $f(i, k)$ be the expected value of the number of cells with size i when the aggregate size of all cells is k. Then, for $i = 1$ we have

$$f(1, k + 1) - f(1, k) = \alpha - (1 - \alpha)f(1, k)/k,\tag{10}$$

where α is the probability that $f(1, k)$ is increased by one and

4. Interpretation of the stochastic processes

Comparing the two processes described by eqs. (15) and (29), respectively, we see that they have identical terms for the addition of new stars to the cells, but that (29), leading to the geometric distribution, has two terms describing the splitting of cells that are absent from (15), leading to the Pareto distribution. These additional terms account for the less skewed shape of the former distribution as compared with the latter, for they amount to a death process that increases frequencies for small i and decreases frequencies for large i.

The simpler process of (15) is easy to interpret as an explanation of city size distributions. It postulates that the population increase, through the net excess of births over deaths and through migration from rural or foreign areas is proportional to current city size. Since these are plausible assumptions under many conditions, it is not surprising that the observed distributions often fit the Pareto Law.

The additional terms in the process of (29) have no easy interpretation in terms of processes of city growth. Cities do not usually split; although, rarely, a standard metropolitan area, as defined by the U.S. Census, will be divided into two such areas. But certainly all possible splits – into various pairs of equal and unequal fragments – do not occur with equal frequency. Thus, the derivation of the geometric distribution from Bose–Einstein statistics, although usual in applications in physics, seems not to be relevant to city sizes,[1] nor does the geometric distribution appear to fit the observed data.

---

[1]This final statement needs to be interpreted carefully. It does *not* assert that the geometric distribution may not be derived from other models approximately obeying Gibrat's law. Indeed Haran and Vining (1973) have published such a derivation, whose assumptions can be given a reasonable interpretation in terms of migration processes. Their derivation, however, is not based on Bose–Einstein statistics.

CHAPTER 5

Some Monte Carlo estimates
of the Yule distribution*

In previous chapters it has been seen that only the simplest stochastic models that embody the Gibrat assumption or some approximation to it are analytically tractible. In particular, it has been essential to make the assumption of a constant rate of entry of new units, although some heuristic analysis was carried out in ch. 1 of the situation where the rate of entry decreases gradually.

Computer simulation methods enable us to go beyond the analytically solvable cases of the stochastic models, and to examine in some detail how their behavior changes as we depart from the limiting assumptions of those cases.

The simulations described in this chapter, carried out by Simon and Van Wormer, were designed specifically to examine the fit of the stochastic models to word frequency data under relaxation of the assumption of a constant rate of entry of new words. The results have interest, however, that goes beyond the specific application to word frequencies. For they show that the slight concavity to the origin that is characteristic of most of the empirical log–log distributions, the firm size distributions as well as the others, can be accounted for by postulating a decreasing

*Any views expressed in this paper are those of the authors. They should not be interpreted as reflecting the views of The RAND Corporation or the official opinion or policy of any of its governmental or private research sponsors.

We are indebted to M. I. Bernstein of The RAND Corporation who wrote the SCAT program for the IBM 7090 version of the Monte Carlo Program reported here.

rate of entry of new units. The change in rate of entry must be rather substantial to produce moderate departures from linearity of the distribution plotted on a log–log scale.

1. Introduction

The stochastic process known as the Yule process and some of its variants have been advanced as explanations for the well-known regularities that are observed in the frequency distributions of words in text (chs. 1 and 2, above). The Yule distribution can be obtained in closed form, by solving the differential or difference equations of the stochastic process, in the special cases where the rate at which new words enter the text is a constant. Approximate solutions have been obtained for a few other cases, but little or nothing has been known of the goodness of the approximations.

The Yule process can be defined in a number of different ways. As applied to word frequency data, the following set of assumptions provides a convenient definition:

Assumption I.   The probability that the $(k + 1)$st word added to a text will be a word that has not occurred before is $\alpha(k)$.

Assumption II.   The probability that the $(k + 1)$st word added to a text will be a word that has already occurred i times $(i \geqslant 1)$ is proportional to $if(i, k)$, where $f(i, k)$ is the number of distinct words (types) that have occurred exactly i times each in the first k words of text.

In the special case where $\alpha(k) = \bar{\alpha}$, a constant, the equations for the process are satisfied by the steady-state solution:

$$f(i) = \rho B(i, \rho + 1), \tag{1}$$

where $B(i, \rho + 1)$ is the beta function of $i, \rho + 1$:

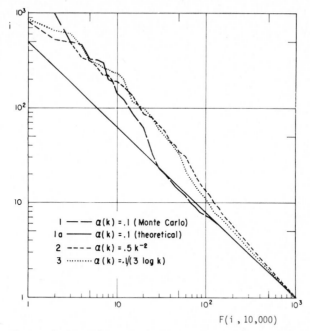

Fig. 5.1. Monte Carlo estimates of the cumulative frequency distribution (average of alpha close to 0.1).

$$\rho_v = \frac{\alpha(k)k}{n(k)} \times \frac{1}{(1 - \alpha(k))}. \tag{4}$$

Columns 1a and 6a of table 5.1 are obtained from the analytic solution of the Yule process for the case where alpha is constant, and can be compared with the Monte Carlo data in columns 1 and 6, respectively. There is close agreement between the theoretical and simulated frequencies.

From the table and the accompanying figure, we can draw a number of conclusions about the Yule distribution with constant or slowly decreasing alpha. In all six cases, the cumulative distributions are very nearly linear on a logarithmic scale, and their slopes are close to unity. In the four cases with decreasing alpha, the distributions are slightly concave downward, and the estimated values of rho for small i are lower than the corresponding theoretical values for constant α. For, taking $\bar{\alpha} =$

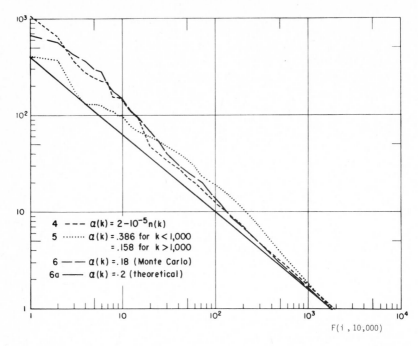

Fig. 5.2. Monte Carlo estimates of the cumulative frequency distribution (average of alpha close to 0.2).

$n(k)/k$, we would have a ρ of approximately 1.11 in the first three cases, and approximately 1.22 in the last three. The values of ρ_2 estimated from the cumulative distribution are very close to the estimate ρ_v. In the case of the function of column 2, $\alpha(k) = \alpha k^{-b}$, Mandelbrot (1959) has proposed $\rho_m = (1 - b)$ as an approximate estimate of ρ. This would give a value of 0.8 for column 2, which is much smaller than the values estimated from the cumulative distribution.

If we can generalize from these six examples (and we have made a substantial number of other simulation runs that give qualitatively similar results), we may say that the distributions in the decreasing-alpha case approximate the distributions in the constant-alpha cases, but deviate slightly in two respects: they are slightly concave downward, and they give slightly smaller rhos for small and moderate i. Both of these deviations from the limiting case have been commonly observed in the empirical data on word distributions, and agree with the conclusions reached by approximate methods in ch. 1.

3. Simulations with alpha estimated from empirical distributions

A more direct test of the ability of the variable-alpha assumption to explain empirical word distribution data is afforded by two samples of text for which we have not only the frequency distributions but also the empirical values of $n(k)$ – from which we can approximate $\alpha(k)$ – for various values of k. One of these is a piece of continuous prose, 10,557 words in length, written by a schizophrenic, Jackson M.; the other is a very large sample of text, from Russian physics journals, 234 096 words in length.[1] In table 5.2 and figs. 5.3 and 5.4, we compare each of these sets of empirical data with three sets of data from Monte Carlo simulations.

The columns in table 5.2 correspond to the following empirical data and simulations, the latter using initial conditions B:

(1) empirical data from writings of Jackson M.,
(2) simulation with constant alpha ($\alpha = 0.18$),
(3) simulation with alpha estimated from empirical data,
(4) simulation taking $\alpha = 0.5$ for $k \leqslant 100$ and $\alpha = 0.179$ for $k > 100$,

[1] The Jackson M. word count was furnished to us by A. W. Dickinson and J. R. Parks, Monsanto Chemical Company, St. Louis. The Russian word count was compiled in connection with research on mechanical translation at The RAND Corporation under the direction of David R. Hayes. In both instances we are grateful for their kindness in giving us access to the aggregate distributions and to data from which $\alpha(k)$ could be estimated for various values of k. These latter data are given in table 5.3. Since $\alpha(k)$ for the Jackson M. sample is approximately $Ak^{-0.3}$ and for the Russian sample, approximately $Bk^{-0.37}$, we obtain estimates for ρ_m of 0.7 and 0.63, respectively. As in the previous cases, these estimates are much too small. The estimates of ρ_v from the empirical $\alpha(k)$ are 0.90 and 0.82 respectively, which agree well with ρ_2 in columns 3 and 7, respectively.

Table 5.2
Comparison of simulations with two empirical word distributions.
(columns are identified in text)

i	1	2	3	4	5	6	7	8
				$f(i, k)$				
1	1 049	1 031	849	1 075	10 171	12 388	10 126	12 385
2	281	338	297	315	3 617	3 933	3 700	3 935
3	153	141	149	153	1 949	1 958	2 160	1 966
4	85	74	103	84	1 233	1 133	1 323	1 132
5	54	62	78	57	821	707	921	708
6	38	30	48	47	622	516	716	518
7	31	34	37	28	547	353	591	357
8	24	21	32	24	403	302	504	300
9	16	15	31	14	346	223	411	222
10	19	7	17	15	276	180	334	182
				$F(i, k)$				
1	1 898	1 883	1 864	1 946	23 081	23 496	24 020	23 782
4	415	373	569	403	7 344	5 217	8 034	5 496
10	167	137	240	149	3 372	1 983	3 568	2 259
40	33	32	40	41	898	421	706	657
100	17	13	9	13	311	158	247	335
400	1	3	0	1	40	45	49	92
1 000	0	0		0	14	21	17	21
4 000					2	8	4	3
10 000					0	2	0	0
i_{max}	623	719	285	525	9 868	25 922	9 241	9 443
ρ_2	1.16	1.14	0.88	1.16	0.84	1.08	0.79	1.06
ρ_{10}	1.06	1.14	0.89	1.12	0.84	1.07	0.83	1.02
ρ_{100}	1.02	1.08	1.16	1.09	0.94	1.09	0.99	0.93
ρ_v		1.22	0.90	1.22		1.11	0.82	1.11

(5) empirical data from Russian sample,
(6) simulation with constant alpha ($\alpha = 0.1$),
(7) simulation with alpha estimated from empirical data,
(8) simulation taking $\alpha = 0.386$ for $k < 1\,000$ and $\alpha = 0.1$ for $k > 1\,000$.

We observe that the frequencies in columns 2, 3, and 4 agree rather well with those in column 1; and the frequencies in columns 6, 7, and 8 with those in column 5. The simulation with alpha estimated from the empirical data (column 7) matches the

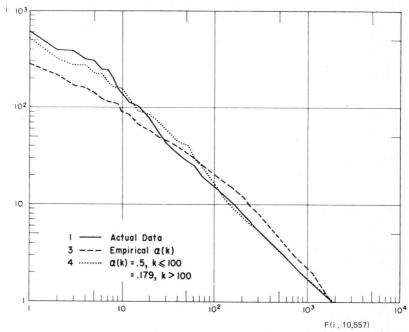

F(i , 10,557)

Fig. 5.3. Cumulative frequency distributions, actual data and Monte Carlo
estimates, Jackson M. data.

data for the Russian sample exceptionally well. On the other
hand, column 4 matches the Jackson M. data better than column
3. We also get a good fit to the Jackson M. data with $\alpha(k) =$
$1/(1.62k)$.

The rationale for the alphas in columns 4 and 8 is that certain
common "functional" words (particularly conjunctions and pre-
positions) come into all the samples at a very early stage – their
number being a hundred or two. Thus, alpha is initially quite
high, drops off rapidly, and then maintains itself at a relatively
constant low level. In ch. 1 it was argued that the·
"contagion" underlying the word distributions involves both
imitative and associative processes. Insofar as the latter are
concerned, the relevant $\alpha(k)$ would be the function actually
estimated from the prose sequence itself. However, if the
introduction of new words depends partially (as it certainly

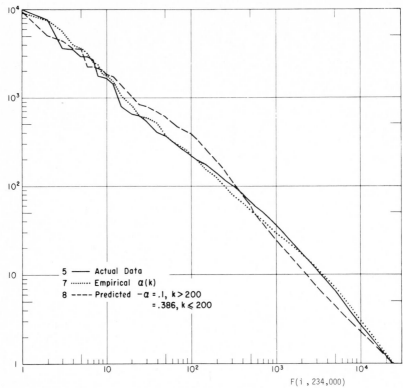

Fig. 5.4. Cumulative frequency distributions, actual data and Monte Carlo estimates, Russian data.

does) on the imitation of word sequences from outside the sample, it is less clear that the function estimated empirically from the sample is the relevant one for the Monte Carlo simulation. We can offer no particularly convincing explanation of why the empirically estimated function gave the better fit in the case of the Russian data but not with Jackson M.'s text. Both simulations give remarkably good fits in both instances.

In speaking of "goodness of fit" we have not invoked any formal statistical tests. Since the hypothesis under consideration – that the data are generated by the Yule process – is an extreme hypothesis, it is not obvious what test would be appropriate. For this reason, we have presented as much of the raw data as feasible, to permit the reader to decide whether he is satisfied with the fits. Our own conclusion is that the Monte Carlo simulations of the Yule process provide rather convincing explanations of the empirical word distributions.

Table 5.3
Rate of entry of new words [$\alpha(k)$].

Jackson M. sample		Russian sample	
k	$\alpha(k)$	$k \times 10^{-3}$	$\alpha(k)$
0–1 000	0.386	0–26.6	0.217
1 001–2 000	0.217	26.6–45.9	0.160
2 001–3 000	0.160	45.9–84.0	0.089
3 001–4 000	0.204	84.0–109.4	0.093
4 001–5 000	0.160	109.4–134.2	0.089
5 001–10 557	0.139	134.2–160.6	0.075
		160.6–186.8	0.065
		186.8–213.4	0.072
		213.4–234.1	0.078

4. Conclusions

In this paper we have presented some data on Monte Carlo simulations of the Yule process under assumptions that the rate at which new words are introduced into the text is slowly decreasing, rather than constant. We have seen that the data deviate slightly in the predicted direction from the theoretical values computed earlier for the constant-alpha case. Finally, we have used the simulation technique to obtain good fits to two empirical word distributions – in the one case using as boundary conditions the empirically observed rate at which new words came into the text. Our results lend additional credence to the hypothesis that the observed gross statistical properties of word frequency distributions are a reflection and consequence of the fundamental processes of imitation and association that are involved in the production of writing and speech.

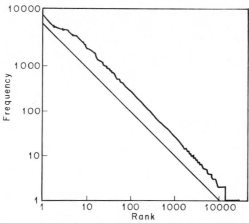

Fig. 6.2. Words occurring in Joyce's *Ulysses* (ranked by frequency of occurrence).

many, 1 017, over 10 000 population. Actually, there were 1 072. It would assert that there were one tenth as many, 203, over 50 000 population; actually, there were 198. It would assert that the largest city, New York, had a population just over ten million people; actually, its population was seven and one half million. The other "facts" asserted above, for cities and words, hold only to comparable degrees of approximation.

At the very least, one would think, the statements of fact should be amended to read "nearly inversely proportional" or "approximately inversely proportional" rather than simply "inversely proportional". But how near is "nearly", and how approximate is "approximately"? What degree of deviation from the bald generalization permits us to speak of an approximation to the generalization rather than its disconfirmation? And why do we prefer the simple but approximate rule to the particular but exact facts?

2. Approximation

It is well known – at least among mathematical statisticians – that the theory of statistical tests gives us no real help in choosing between an approximate generalization and an invalid one.[1] By imbedding our generalization in a probability model, we can ask: If this model describes the real "facts" what is the probability that data would have occurred at least as deviant from the generalization as those actually observed? If this probability is very low – below the magic one percent level, say – we are still left with two alternatives: the generalization has been disconfirmed, and is invalid; or the generalization represents only a first approximation to the true, or "exact" state of affairs.

Now such approximations abound in physics. Given adequate apparatus, any student in the college laboratory can "disconfirm" Boyle's law – i.e. can show that the deviations of the actual data from the generalization that the product of pressure by volume is a constant are too great to be dismissed as "chance." He can "disconfirm" Galileo's law of falling bodies even more dramatically – the most obvious way being to use a feather as the falling body.

When a physicist finds that the "facts" summarized by a simple, powerful generalization do not fit the data exactly, his first reaction is *not* to throw away the generalization, or even to complicate it by incorporating additional terms. When the data depart from $s = \frac{1}{2}gt^2$, the physicist is not usually tempted to add a cubic term to the equation. (It took Kepler almost ten years to retreat from the "simplicity" of a circle to the "complexity" of an ellipse.) Instead, his explorations tend to move in two directions: (1) toward investigations of his measurement proce-

---

[1]For a brief, but adequate statement of the reasons why "literally to test such hypotheses . . . is preposterous", see Savage (1954, pp. 254–256). Since such tests are still reported frequently in the literature, it is perhaps worth quoting Savage (1954, p. 254) at slightly greater length: "The unacceptability of extreme null hypotheses is perfectly well known; it is closely related to the oftenheard maxim that science disproves, but never proves, hypotheses. The role of extreme hypotheses in science and other statistical activities seems to be important but obscure. In particular, though I, like everyone who practices statistics, have often 'tested' extreme hypotheses, I cannot give a very satisfactory analysis of the process, nor say clearly how it is related to testing as defined in this chapter and other theoretical discussions."

dures as possible sources of the discrepancies; and (2) toward the identification of other variables associated with the deviations. These two directions of inquiry may, of course, be interrelated.

In his concern with other variables, the physicist is not merely or mainly concerned with "control" in the usual sense of the term. No amount of control of air pressure, holding it, say, exactly at one atmosphere, will cause a feather to obey Galileo's law. What the physicist must learn through his explorations is that as he decreases the air pressure on the falling body, the deviations from the law decrease in magnitude, and that if he can produce a sufficiently good vacuum, even a feather can be made to obey the law to a tolerable approximation.

In the process of producing conditions under which deviations from a generalization are small, the scope of the generalization is narrowed. Now it is only claimed to describe the facts "for an ideal gas," or "in a perfect vacuum." At best, it is asserted that the deviations will go to zero in the limit as the deviation of the actual experimental conditions from the "ideal" or "perfect" conditions goes to zero.

At the same time that the breadth of the empirical generalization is narrowed by stating the conditions, or limiting conditions, under which it is supposed to hold, its vulnerability to falsification is reduced correspondingly. Since this is a familiar feature of theorizing in science, we will not elaborate on the point here.

Occasionally, an empirical generalization is abandoned, after innumerable attempts to tidy it up have failed. Bode's law, that the successive distances of the planets from the sun constitute an approximate geometric series, is an example of a regularity now regarded as perhaps "accidental," through failure to discover limiting conditions that would regularize it, or underlying processes that would account for it. Newton's laws are *not* an example, for they were saved (a) by limiting them to conditions where velocities are low relative to the velocity of light, and (b) by showing that just under those conditions they can be derived in the limit from the more general laws of relativity.

From these, and many other examples, we can see what

importance the physical and biological sciences attach to finding simple generalizations that will describe data approximately under some set of limiting conditions. Mendel's treatment of his sweet-pea data, as reflecting simple ratios of 3 to 1 in the second-generation hybrids, is another celebrated illustration; as is Prout's hypothesis (uneasily rejected by chemists for several generations until its exceptions were explained by the discovery of isotopes) that all atomic weights are integral multiples of the weight of the hydrogen atom. All of these examples give evidence of strong beliefs that when nature behaves in some unique fashion – deals a hand of thirteen spades, so to speak – this uniqueness, even if approximate, cannot be accidental, but must reveal underlying lawfulness.

3. The law-finding process

Let us return to city sizes and word frequencies. We have described the law-finding process in two stages:

(1) finding simple generalizations that describe the facts to some degree of approximation;
(2) finding limiting conditions under which the deviations of facts from generalization might be expected to decrease.

The process of inference from the facts (the process called "retroduction" by Peirce and Hanson[2]) does not usually stop with this second stage, but continues to a third:

(3) explaining why the generalization "should" fit the facts. (Examples are the statistical-mechanical explanation for Boyle's law or Boyle's own "spring of the air" explanation, and Newton's gravitational explanation for Galileo's law.)

Before we go on to this third stage, we must consider whether we have really been successful in carrying out the first two for the rank-size distributions.

[2]Hanson (1961, pp. 85–88).

Does the generalization that size varies inversely with rank really fit the facts of cities and words even approximately? We plot the data on double log paper. If the generalization fits the facts, the resulting array of points will (1) fall on a straight line, (2) with a slope of minus one.

Since we earlier rejected the standard statistical tests of hypotheses as inappropriate to this situation, we are left with only judgmental processes for deciding whether the data fall on a straight line. It is not true, as is sometimes suggested, that almost *any* ranked data will fall on a straight line when graphed on doubly logarithmic paper. It is quite easy to find data that are quite curvilinear to the naked eye (see fig. 6.3). Since we are not committed to exact linearity but only approximate linearity, however, the conditions we are imposing on the data are quite weak, and the fact that they meet the conditions is correspondingly unimpressive. We may therefore find the evidence unconvincing that the phenomena are "really" linear in the limiting cases. The phenomena are not striking enough in this respect to rule out coincidence and chance. Should we believe the data to be patterned?

It has often been demonstrated in the psychological laboratory that men – and even pigeons – can be made to imagine patterns in stimuli which the experimenter has carefully constructed by random processes. This behavior is sometimes called "superstitious," because it finds causal connections where the experimenter knows none exist in fact. A less pejorative term for such behavior is "regularity-seeking" or "law-seeking." It can be given a quite respectable Bayesian justification. As Jeffreys and Wrinch (1921) have shown, if one attaches a high a priori probability to the hypothesis that the world is simple (i.e. that the facts of the world, properly viewed, are susceptible to simple summarization and interpretation); and if one assumes also that simple configurations of data are sparsely distributed among all logically possible configurations of data, then a high posterior probability must be placed on the hypothesis that data which appear relatively linear in fact reflect approximations to conditions under which a linear law holds.

The reason that apparent linearity, by itself, does not impress us is that it does not meet the second condition assumed above – the sparsity of simple configurations. A quadratic law, or an exponential, or a logarithmic, are almost as simple as a linear one; and the data they would produce are not always distinguishable from data produced by the latter.

What is striking about the city size and vocabulary data, however, is not just the linearity, but that the slope of the ranked data, on a log scale, is very close to minus one. Why this particular value, chosen from the whole nondenumerable infinity of alternative values? We can tolerate even sizeable deviations from this exact slope without losing our confidence that it must surely be the limiting slope for the data under some "ideal" or "perfect" conditions.

We might try to discover these limiting conditions empirically, or we might seek clues to them by constructing an explanatory model for the limiting generalization – the linear array with slope of minus one. In this way we combine stages two and three of the inference process described at the beginning of this section. Let us take this route, confining our discussion to city size distributions.

4. Explaining regularity

To "explain" an empirical regularity is to discover a set of simple mechanisms that would produce the former in any system governed by the latter. A half dozen sets of mechanisms are known today that are capable of producing the linear rank-size distribution of city populations. Since they are all variations on one or two themes, we will sketch just one of them (ch. 1).

We consider a geographical area that has some urban communities as well as rural population. We assume, for the urban population, that birth rates and death rates are uncorrelated with city size. ("Rate" here always means "number per year per 1 000 population.") We assume that there is migration between cities, and net emigration from rural areas to cities (in addition to net immigration to cities from abroad, if we please). With respect to all migration, we assume: (1) that out-migration rates from cities are uncorrelated with city size; (2) that the probability that any migrant, chosen at random, will migrate to a city in a particular size class is proportional to total urban population in that class of cities. Finally, we assume that of the total growth of population in cities above some specified minimum size, a constant fraction is contributed by the appearance of new cities (i.e. cities newly grown to that size). The resulting steady-state rank-size distribution of cities will be approximately linear on a double log scale, and the slope of the array will approach closer to minus one as the fraction of urban population growth contributed by new cities approaches zero.

When we have satisfied ourselves of the "reasonableness" of the assumptions incorporated in our mechanism, and of the insensitivity of the steady-state distribution to slight deviations from the assumptions as given, then we may feel, first, that the empirical generalization can now be regarded as "fact;" and, second, that it is not merely "brute fact" but possesses a plausible explanation.

But the explanation does even more for us; for it also suggests under what conditions the linearity of the relation should hold most exactly, and under what conditions the slope should most closely approximate to one. If the model is correct, then the rank-size law should be best approximated in geographical areas (1) where urban growth occurs largely in existing cities, (2) where all cities are receiving migration from a common "pool;" and (3) where there is considerable, and relatively free, migration among all the cities. The United States, for example, would be an appropriate area to fit the assumptions of the model; India a less suitable area (because of the relatively weak connection between its major regions); Austria after World War I a still less suitable area (because of the fragmentation of the previous Austro–Hungarian Empire, see fig. 6.3). We do not wish to discuss the data here beyond observing that these inferences from the model seem generally to be borne out.

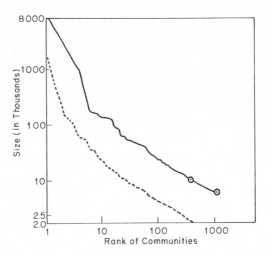

Fig. 6.3. Rank-size distribution of cities in Austro-Hungarian Empire, 1910
(———) and in Austria, 1934 (– – – –).

5. Simplicity

In our account thus far, the simplicity of the empirical generalization has played a central role. Simplicity is also an important concept in Popper (1961)[3] but Popper treats simplicity in a somewhat different way than we have done. Popper (p. 140) equates simplicity with *degree of falsifiability*. A hypothesis is falsifiable to the degree that it selects out from the set of all possible worlds a very small subset, and asserts that the real world belongs to this subset.

There is a strong correlation between our intuitive notions of simplicity (e.g. that a linear relation is simpler than a polynomial of higher degree) and falsifiability. Thus, among all possible monotonic arrays, linear arrays are very rare. (They would be of measure zero, if we imposed an appropriate probability measure on the set of all monotonic arrays.) Linear arrays with slope of minus one are even rarer.

[3]Especially ch. VII.

No one has provided a satisfactory general measure of the simplicity or falsifiability of hypotheses. In simple cases, the concepts have an obvious connection with degrees of freedom: the fewer the degrees of freedom, or free parameters, the simpler and more falsifiable the hypothesis. We shall not undertake to carry the formalization of the concepts beyond this intuitively appealing basis.[4]

Notice, however, that our use of simplicity is quite different from Popper's (1961). Popper's argument runs like this: it is desirable that hypotheses be simple so that, if they are false, they can be disconfirmed by empirical data as readily as possible. Our argument (apparently first introduced by Jeffreys and Wrinch (1921)) runs: a simple hypothesis that fits data to a reasonable approximation should be entertained, for it probably reveals an underlying law of nature. As Popper himself observes (Popper (1961, p. 142, fn. 2)), these two arguments take quite opposite positions with respect to the "probability" or "plausibility" of simple hypotheses. He regards such hypotheses as describing highly particular, hence improbable states of the world, and therefore as readily falsified. Jeffreys and Wrinch (1921) (and we) regard them as successfully summarizing highly unique (but actual) states of the world, therefore as highly plausible.

Which of these views is tenable would seem to depend on which came first, the generalization or the data. If we construct generalizations, with no criterion to guide our choice except that they be simple, and subsequently apply them to data, then the simpler the generalization the more specific their description, and the less likely that they will stand up under their first empirical test. This is essentially Popper's argument.

But the argument does not apply if the generalization was constructed with the data in view. The rank-size hypothesis arises because we think to plot the data on double log paper, and

---

[4]The most serious attempts at formalization are those undertaken by Jeffreys and Wrinch (1921) and Goodman (1958). We must note in passing that in his discussion of the former authors Popper (1961) does not do justice to their technical proposal for introducing prior probabilities based on simplicity.

when we do, it appears to be linear and to have a slope of minus one. There is no thought of using the data to falsify the generalization, for the latter has come into being only because it fits the data, at least approximately.

Now one can cite examples from the history of science of both of these alternative sequences of events. It is probably true, however, that the first sequence – generalization followed by data – seldom occurs except as a sequel to the second. The special theory of relativity, for example, led to the prediction of the convertibility of mass into energy. But special relativity itself was based on a generalization, the Lorentz–Fitzgerald equation, that was derived to fit facts about the behavior of particles in very intense fields of force, as well as other facts about electromagnetics and the "luminiferous ether." Special relativity did not commend itself to Einstein merely because of its "simplicity" independently of the facts to be explained (the Galilean transformations would be thought by most people to be simpler than the Lorentz).

If the generalization is just that – an approximate summary of the data – then it is certainly not falsifiable. It becomes falsifiable, or testable, when (a) it is extended beyond the data from which it was generated, or (b) an explanatory theory is constructed, from which the generalization can be derived, and the explanatory theory has testable consequences beyond the original data.

With respect to the city size data, case (a) would arise if the rank-size generalization were proposed after examining the data from the 1940 U.S. Census, and then were extrapolated to earlier and later dates, or to the cities of other countries. Case (b) would arise if we were to note that the explanatory theory of §4, above, has implications for patterns of migration that could be tested directly if data on points of origin and destination of migrants were available.

It should be evident that the mechanisms incorporated in the explanatory theory were not motivated by their falsifiability. They were introduced in order to provide "plausible" premises from which the generalization summarizing the observed data could be deduced. And what does "plausible" mean in this context? It means that the assumptions about birth and death rates and migration are not inconsistent with our everyday general knowledge of these matters. At the moment they are introduced, they are already known (or strongly suspected) to be not far from the truth. The state of affairs they describe is not rare or surprising (given what we actually know about the world); rather their subsequent empirical falsification would be rather surprising. What is *not* known at the moment they are introduced is whether they provide adequate premises for the derivative of the rank-size generalization.

Explaining the empirical generalization, that is, providing a set of mechanisms capable of producing it, therefore reintroduces new forms of testability to replace those that were lost by accepting the approximation to the data. Even without data on migration, the mechanism proposed to explain the city rank-size law can be subjected to new tests by constructing the transition matrix that compares the sizes of the same cities at two points of time (taking the 1900 population, say, as the abscissa, and the 1950 population as the ordinate (see fig. 6.4)). The explanatory mechanism implies that the means of the rows in this matrix fall

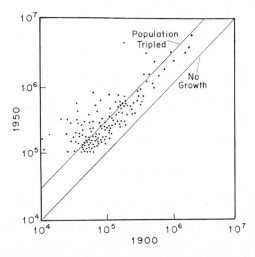

Fig. 6.4. Population of U.S. metropolitan districts, 1900 and 1950 (only districts over 100 000 population in 1950 are shown).

on a straight line through the origin (or on a straight line of slope
+1 on a log–log scale). The result (which we will expect to hold
only approximately) is equivalent to the proposition that the
expected growth rates are independent of initial city size.

6. An example from psychology

In the preceding sections a model has been sketched of the
scientific activities of hypothesis-generation and hypothesis-
testing. The model suggests that there are several distinct
processes that lead to hypotheses being formulated, judged with
respect to plausibility, and tested. One of these processes, the
generation of simple extreme hypotheses from the "striking"
characteristics of empirical data, fits closely the idea of Jeffreys
and Wrinch (1921) that simple hypotheses possess a high plausibility. A second process, the construction of explanations for
these extreme hypotheses, takes us back to Popper's (1961) idea
that simple hypotheses entail strong and "improbable" consequences, hence are readily falsified (if false). There is no contradic-
tion between these two views.

To elucidate further this model of the scientific process, and to
reveal some additional characteristics it possesses, the remaining sections of this paper will be devoted to the analysis of a
second example, this one of considerable interest to the psychology of learning and concept formation. An important question in
psychology during the past decade has been whether learning is
to be regarded as a sudden, all-or-none phenomenon, or whether
it is gradual and incremental. One value in stating the question
this way is that the all-or-none hypothesis is a simple, extreme
hypothesis, hence is highly falsifiable in the sense of Popper
(1961).

The experiments of Rock (1957) first brought the all-or-none
hypothesis into intense controversy. His data strongly supported
the hypothesis (even under rather strict limits on the degree of
approximation allowed). Since his generalization challenged
widely-accepted incrementalist theories, his experiment was

soon replicated (seldom quite literally), with widely varying findings. The discussion in the literature, during the first few years after Rock's initial publication, centered on the "validity" of his data – i.e. whether he had measured the right things in his experiment, and whether he had measured them with adequate precision.

Only after several years of debate and publication of apparently contradictory findings was some degree of agreement reached on appropriate designs for testing the hypothesis. Still, some experimenters continued to find one-trial learning, others incremental learning. After several more years, the right question was asked, and the experiments already performed were reviewed to see what answer they gave.[5] The "right question," of course, was: "Under what conditions will learning have an all-or-none character?" The answer, reasonably conformable to the experimental data, commends itself to common sense. Oversimplified, the answer is that one-trial learning is likely to occur when the time per trial is relatively long, and when the items to be learned (i.e. associated) are already familiar units.[6] There are the "ideal" or "perfect" conditions under which one-trial learning can be expected to occur.

7. Improving the generalization

Meanwhile, the all-or-none hypothesis was also being applied to concept attainment experiments. Important work was done in

[5]Postman (1963) and Underwood and Keppel (1964).

[6]As a matter of history, we might mention that in 1957, prior to Rock's (1957) publication of his experiment, a theory of rote learning, designed especially to explain data that were in the literature prior to World War II (the serial position curve, the constancy of learning time per item, some of E. Gibson's experiments on stimulus similarity) had been developed by E. Feigenbaum and H. Simon. This theory, EPAM, was sufficiently strong to predict the conditions under which one-trial learning would occur. It was not widely known among psychologists at that time, however, and had little immediate influence on the controversy. (But see Gregg, Chenzoff and Laughery (1963) and Gregg and Simon (1967b).)

this area by Estes, by Bourne, and by Bower and Trabasso, among others. We will take as our example for discussion a well-known paper by Bower and Trabasso that Gregg and Simon have analysed in another context.[7]

The experiments we shall consider employ an N-dimensional stimulus with two possible values on each dimension, and having a single relevant dimension (i.e. simple concepts). On each trial, an instance (positive or negative) is presented to the subject; he responds "positive" or "negative;" and he is reinforced by "right" or "wrong," as the case may be.

Bower and Trabasso obtain from the data of certain of their experiments an important empirical generalization: the probability that a subject will make a correct response on any trial prior to the trial on which he makes his last error is a constant. (In their data, this constant is always very close to one half, but they do not incorporate this fact in their generalization as they usually state it.) Since the generalization that the probability of making a correct response is constant is an extreme hypothesis, the standard tests of significance are irrelevant. We must judge whether the data fit the generalization "well enough". Most observers, looking at the data, would agree that they do (see fig. 6.5).

But Bower and Trabasso go a step further. They derive the empirical generalization from a simple stochastic model of the learning process – they explain it, in the sense in which we used that term earlier. The explanation runs thus: (1) the subject tries out various hypotheses as to what is the correct concept, and responds on individual trials according to the concept he is currently holding; (2) if his response is wrong, he tries a new concept. Two important empirical quantities are associated with the model: the probability of making a correct response prior to the last error; and the probability that any particular trial will be the trial of last error.

Now there are in fact *two* distinct all-or-none generalizations that can be formulated in terms of these two empirical quan-

[7]Bower and Trabasso (1964); Gregg and Simon (1967a).

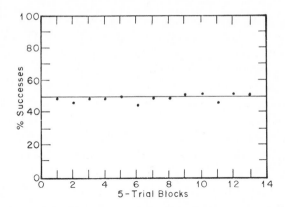

Fig. 6.5. Concept experiment: percentage of successes prior to the last error (from Bower and Trabasso).

tities. The first, already mentioned, is the generalization that the probability of making a *correct response* is constant as long as the subject holds the wrong hypothesis about the concept (i.e. up to the trial of his last error). The second, quite different, is the generalization that the probability of switching to the *correct hypothesis* about the concept does not change over trials (i.e. that the probability is constant that each trial will be the trial of last error).

To test the first (correct response) all-or-none generalization, we have one datum from each subject for each trial prior to his last error – a considerable body of data to judge the approximation of the error rate to a constant. To test the second (correct hypothesis) all-or-none generalization, we have only one datum from each subject – the trial on which he made his last error. Hence, trial-to-trial changes in the probability of switching to the right concept are confounded with differences in that probability among subjects. If, for any single subject, this probability increases with trials, the increase is counterbalanced by the fact that the subjects with lowest average probability will tend to learn last. Thus (as Bower and Trabasso are careful to point out) the data to test the second generalization directly are scanty and inadequate.

8. Stochastic models and process models

The Bower–Trabasso stochastic model is an explanation of the observed constancy of the error rate. But it is a very bland model, making rather minimal assumptions about the process that is going on. We can pursue the goal of explanation a step further by constructing a more detailed model of the cognitive processes used by subjects in concept attainment, then using this detailed model to subject the theory to further tests. (As Gregg and Simon have shown in a paper on this topic (Gregg and Simon (1967a)), Bower and Trabasso do, in fact, employ such a process model, but only informally.)

There are two important differences between the summary stochastic model and the more detailed process model. The process model, but not the stochastic model, spells out how the experimenter selects (on a random basis) the successive instances, how the subject responds, and how he selects a new concept when his current one is found wrong. The stochastic model, but not the detailed model, contains two free parameters, one specifying the probability that the subject's response will be (fortuitously) correct when he does not hold the correct concept; the other specifying the probability that he will select the correct concept as his new one when his current concept is found wrong.

The stochastic model and process model can be formalized by stating them in a computer programming language (Gregg and Simon (1967a)). When this is done, it is found that the stochastic model requires 15 statements – i.e. simple computer instructions – for its formulation, the detailed process model 27. Against this parsimony of the stochastic model must be balanced the fact that that model contains two free numerical parameters, the process model none. Which model is the simpler?

If we apply Popper's criteria of simplicity – the simpler theory being the one that is more highly falsifiable – then the question has a definite answer. The detailed process model is simpler than the stochastic model (see Gregg and Simon (1967a, pp. 271–272)). For, by a straightforward aggregation of variables, the stochastic model, with particular values for the free parameters, can be derived deductively from the process model. Hence, the process model is a special case of the stochastic model. (The process model predicts an error rate of about 0.5 per trial prior to the trial of last error. It also predicts the probability that the last error will occur on a particular trial, but this probability depends on the structure of the stimuli – the number of attributes they possess, and the number of values of each attribute.)

The additional detail incorporated in the process model's assumptions also provides additional opportunities for subjecting the model to empirical test. The hypotheses held by the subject as to the correct concept do not appear explicitly in the stochastic model; hence data relating to these hypotheses (obtained, say, by asking the subject on each trial what concept he holds, as was done by Feldman (1964), or obtained by indirect procedures developed by Levine (1966)) cannot be used to test that model, but can be used to test the process model.

If parsimony refers to the brevity with which a theory can be described, then the stochastic model is the more parsimonious (fifteen statements against twenty-seven). But lack of parsimony, so defined, must not be confused with degrees of freedom. We have seen in this case that the less parsimonious theory is the simpler (by Popper's (1961) criterion), and by far the more falsifiable.

Testing the detailed process theory raises all the problems mentioned earlier with respect to extreme hypotheses. If the error rate on guessing trials deviates from 0.5 should the theory be rejected? How much of a deviation should be tolerated? In how many cases can a subject report he is holding a concept different from that predicted by the theory before we reject the latter? We have given reasons earlier for thinking that these questions are judgmental, and for concluding that the theory of statistical tests offers no help in answering them. A judgmental answer is that the theory should be rejected only if it turns out to be "radically" wrong. Otherwise, deviations should lead to a search for variables to account for them, and for the "ideal" limiting case in which they would disappear.

Justice Holmes once said: "Logic is not the life of the law." We would paraphrase his aphorism by saying: "Statistics is not the life of science." No existing statistical theory explains what scientists do (or should do) to retroduce, develop, test, and modify scientific theories.

9. Interpreting models

Just as statistically significant deviations of data from a generalization should not always, or usually, lead us to abandon the generalization, so we should not be unduly impressed by excellent statistical fits of data to theory. More important than whether the data fit is why they fit – i.e. what components in the theory are critical to the goodness of fit. To answer this question, we must analyse the internal structure of the theory.

For example, under the conditions where all-or-none learning can be expected to take place, the learning trials can generally be divided into two parts: an initial sequence prior to learning, during which the subject can only guess at the correct answer; a terminal sequence, during which the subject knows the correct concept, and makes no new mistakes. Let us suppose that the boundary between these two segments can be detected (as it can in the concept-learning experiments by the trial on which the last error is made).

Under these conditions, no important conclusions can be drawn about psychological characteristics of the subjects by examining the statistical structure of their responses prior to learning. For the statistics of these responses are simply reflections of the experimenter's randomization of the sequence of stimuli. In one experiment, Estes (1956), for example, employed three different conditions differing only with respect to the number of alternative responses (2, 4 and 8, respectively) available to the subject (see Simon (1962)). He found that the relative number of errors per trial made in these three conditions could be represented by the formula, $A(N-1)/N$, where A is a constant and N is the number of alternative responses.

The data on relative numbers of errors fit this formula with great accuracy – a clearcut case of success for an extreme hypothesis of the kind we have been commending in this chapter. However, the hypothesis that was being tested was not a generalization about psychology, but a well-known generalization about the laws of probability: that in drawing balls at random from an urn containing white and black balls in the ratio of 1 to $(N - 1)$, on the average $(N - 1)/N$ of the balls drawn will be black. This is true regardless of whether the subjects themselves, prior to learning, thought they were simply guessing or thought they were responding in selective, patterned ways to the stimuli. By randomizing the sequence of stimuli presented, the experimenter guaranteed the applicability of the laws of probability to the subject's errors, independently of the systematic or "random" character of the subject's behavior.

As we have pointed out elsewhere, a number of other excellent fits of simple generalizations to data can be attributed to the random presentation of stimuli, rather than to characteristics of the subjects (Simon (1957), Simon (1962), Gregg and Simon (1967a)). This does not imply that it is useless to extract the underlying regularities from the data; but we must be careful to provide the regularities with a correct explanation. To do so, we must examine the internal structure of the theories that lead to the successful generalization.

10. Retroduction of hypotheses

Throughout this paper, considerable stress has been placed on the close interaction between hypotheses and data in the building and testing of theories. In most formal theories of induction, particularly those that belong to the genus "hypothetico-deductive" or "H–D," hypotheses spring full-blown from the head of Zeus, then are tested with data that exist, timelessly and quite independently of the hypotheses.[8] Theories as otherwise

---

[8]For a criticism of this view, see Simon (1955a). That paper was concerned specifically with the relative dating of theory and data, and while I still

divergent as Popper's and Carnap's share this common framework.

It was one of Norwood Hanson's important contributions to challenge this separation of hypothesis from data, and to demonstrate that in the history of science the retroduction of generalizations and explanations from data has been one of the central and crucial processes. In making his point, Hanson was careful not to revert to naive Baconian doctrines of induction. To look at a series of rank-size distributions, approximately log-linear with slopes of minus one; then to conclude that *all* such distributions share these properties, is Baconian. To look at the raw data, and conclude that they can be described adequately by the log-linear function with slope of minus one is not Baconian. It is the latter form of derivation of generalizations from data with which Hanson was primarily concerned, and to which he (following Peirce) applied the name "retroduction."

One of our principal theses here has been that hypotheses retroduced in this way are usually highly plausible, and not highly improbable, as Popper (1961) would insist. We have already resolved part of the apparent paradox. The "improbability" to which Popper refers is improbability of the very special state of nature described by the empirical generalization, not improbability of the generalization itself. But it remains to understand how the scientist can ever be lucky enough to discover the very special generalizations that describe these a priori improbable (but actual) states of nature.

Fortunately, considerable light has been cast on this question by progress in the past decade in our understanding of the theory of human problem solving (Simon (1966)). If the scientist had to proceed by searching randomly through the (infinite) space of possible hypotheses, comparing each one with the data until he found one that matched, his task would be hopeless and endless.

---

subscribe to the general position set forth there – that this dating is relevant to the corroboration of hypothes. ; by data – I would want to modify some of my specific conclusions about the form of the relevance, as various paragraphs in the present chapter will show. [H.A.S.]

This he does not need to do. Instead, he extracts information from the data themselves (or the data "cleaned up" to remove some of the noise), and uses this information to construct the hypothesis directly, with a modest amount of search.

Let us consider a concrete example (Banet (1966)). Suppose we are presented with the sequence: $\frac{9}{5}, \frac{4}{3}, \frac{25}{21}, \frac{9}{8}, \ldots$. What simple generalization can we discover to fit this sequence? We note that all the numerators are squares, that the first and third denominators are four less than their numerators, the second and fourth denominators are one less. We notice that the sequence appears to be monotone decreasing, and to approach a limit – perhaps unity. Nine is 3^2, 25 is 5^2. Suppose we number the terms 3, 4, 5, 6. The corresponding squares are 9, 16, 25, 36. Let's multiply numerator and denominator of the second and fourth terms by four, getting: $\frac{9}{5}, \frac{16}{12}, \frac{25}{21}, \frac{36}{32}, \ldots$. Now the empirical generalization is obvious: the general term of the sequence is $n^2/(n^2 - 4)$. Physicists will recognize this as the well-known Balmer series of the hydrogen spectrum, and what we have done is to reconstruct hypothetically part of Balmer's retroduction. (He probably followed a somewhat different path, and we have only considered the last half of his problem of getting from data to generalization, but this partial and somewhat unhistorical example will serve to illustrate our central point. For the actual history, see Banet's (1966) interesting paper.)

However great a feat it was for Balmer to extract his formula from the data, the process he used was certainly not one of generating random hypotheses, then testing them. It is better described as a process of searching for the pattern in the data. It can be shown, for a considerable class of patterns that are of practical importance, in science, in music, and in intelligence tests, that the range of relations the searcher must be prepared to detect is quite small. It may be that these are the sole relations from which the simplicity of nature is built; it may be they are the only relations we are equipped to detect in nature. In either event, most of the patterns that have proved important for science are based, at bottom, on these few simple relations that humans are able to detect.

11. Conclusions

In this paper, we have examined several aspects of the problem of testing theories, and particularly those important theories that take the form of extreme hypotheses. In part, our argument has been aimed at a negative goal – to show that when we look at realistic examples from natural and social science, statistical theory is not of much help in telling us how theories are retroduced or tested.

As an alternative to standard probabilistic and statistical accounts of these matters, we have proposed that we take into account a whole sequence of events:

(1) The enterprise generally begins with empirical data, rather than with a hypothesis out of the blue.

(2) "Striking" features of the data (e.g. that they are linear on a log scale with slope of minus one) provide for a simple generalization that summarizes them – approximately.

(3) We seek for limiting conditions that will improve the approximation by manipulating variables that appear to affect its goodness.

(4) We construct simple mechanisms to explain the simple generalizations – showing that the latter can be deduced from the former.

(5) The explanatory theories generally make predictions that go beyond the simple generalizations in a number of respects, and hence suggest new empirical observations and experiments that allow them to be tested further.

"Testing" theories, as that process is generally conceived, is only one of the minor preoccupations of science. The very process that generates a theory (and particularly a simple generalization) goes a long way toward promising it some measure of validity. For these reasons, histories of science written in terms of the processes that discover patterns in nature would seem closer to the mark than histories that emphasize the search for data to test hypotheses created out of whole cloth.

II
Firm-size distributions

CHAPTER 7

The size distribution of business firms*

With this chapter, written by Simon and Bonini, we turn from the general discussion of the Yule and Pareto distributions to the substantive task of explaining with the aid of these distributions the observed data on sizes of business firms. Here, several tests are applied to see whether the theoretical functions fit data on large American and British firms, as well as some American data on firm sizes in individual industries.

The chapter illustrates a number of ways of testing the theory. The transition matrices of changes in size from an earlier year to a later can be examined for homoscedasticity on a log–log scale. Average rates of growth of small, medium-sized, and large firms can be compared. A direct estimate of the rate of entry of new firms can be compared with the estimate derived from the slope of the Pareto curve. Estimates of minimum feasible plant size can be compared with the lower end of the distribution, to see if a sharp dropoff in frequencies occurs near that point.

The Yule distribution is used as our primary theoretical construct in fitting the firm-size data. This does not imply that other distributions such as the log-normal or Fisher's log series distribution are never applicable. In fact in ch. 8 we use a system leading to the log-normal distribution as a means of examining the degree of autocorrelation in firm-size growth. For most empirical data, the Yule distribution shows a better fit, especially

*This analysis has been aided by a generous grant from the Ford Foundation for research on organizations. We are grateful to B. Wynne, who performed some of the exploratory analysis, and to F. Modigliani, with whom we have had several enlightening discussions about the theory.

in the upper tail, but for some the log-normal distribution appears to fit better (see the discussion and examples in Cohen (1966) and in Steindl (1965, Appendix B)).

However, as we indicated in the Introduction to this book, the appropriateness of a theoretical model is not dictated solely by how well it fits the empirical data. The plausibility of underlying assumptions must also be taken into account.

While the log-normal distribution is obtained from a random walk based on the Gibrat assumption, the elements in the population must start the random walk all at the same time. If the assumptions are modified to allow new entrants after the process has started, generally the log-normality is destroyed. In the case of the Yule process, new entrants are an integral part of the generating stochastic process. Allowance for new entrants seems essential in order to model the observed phenomena.

The stochastic process leading to the Yule distribution does not allow for decreases in the sizes of particular firms. While this also is not an accurate representation of the empirical phenomena (none of the theoretical models are!), it is probably not a serious defect in the model in an economy where growth is the rule and such decreases are infrequent. In ch. 9, where we are viewing the deviations of individual growth rates from industry averages, and where we are concerned with a fixed population of firms, we employ a process that yields the log-normal distribution in equilibrium.

1. Introduction

The distribution of business firms by size has received considerable attention from economists interested in the phenomena of competition and oligopoly and in the issues of government regulation to which these phenomena are relevant. That the size distribution of firms (whether within a single industry or in a whole economy) is almost always highly skewed, and that its upper tail resembles the Pareto distribution has often been observed, but has not been related very much to economic theory. Attempts at economic explanation of the observed facts about concentration of industry have almost always assumed that the basic causal mechanism was the shape of the long-run average cost curve; but there has been little discussion of why this mechanism should produce, even occasionally, the particular highly skewed distributions that are observed.

In §2 we shall discuss the adequacy of explanations of the size distribution based on the static cost curve. In §3 we shall propose an alternative theory based on a stochastic model of the growth process. In §4 we shall examine the empirical data in the light of the model. In §5 we shall examine the implications of our analysis for public policy. In §6 we shall comment on some of the needs for empirical and theoretical research in this area.

2. Economic theory of the size of firms

Economic theory has little to say about the distribution of firm sizes. In general, we are led to expect a U-shaped long-run cost curve or planning curve for a firm. But the scale corresponding to minimum costs need not be the same for different firms, even in the same industry. If we employ the concept of economic rent, we can say that firms will have the same minimum cost, but varying outputs at this cost (intro., pp. 7–10). If this is the case, the cost curve yields no prediction about the distribution of firms by size and no explanation as to why the observed distributions approximate the Pareto distribution.

Some theorizing has been concerned with long-run increasing, decreasing and constant cost curves for firms (Viner (1931, esp. pp. 210–17)). But the theorizers have hesitated to draw conclusions about the observed size distributions. In some cases, the theory is indeterminate about the distribution, as in the case of constant costs (Viner (1931, p. 211)). In others, the theorists point out that "industry" is such a vague and arbitrary term that comparing the sizes of different firms is like comparing oranges and apples. Differences in the size of markets for firms and the idea that firms are moving towards the equilibrium of the cost curve but haven't reached it are also mentioned as reasons why firms widely varying in size can survive in the same industry.

All these factors make static cost theory both irrelevant for understanding the size distributions of firms in the real world and empirically vacuous. And yet these distributions show such a regular and docile conformity to the Pareto distribution that we would expect some mechanism to be at work to account for the observed regularity.

In the previous discussion (as in much of the literature on this topic) our comments about the long-run cost curve have been a priori. However, J. S. Bain (1956) has made a careful analysis of all the available information on the cost curves of firms and plants in a substantial number of industries, using both published and original data that he obtained by questionnaire. His data show that plant cost curves (ignoring the problem of intra-industry specialization) generally are J-shaped. Below some critical scale unit costs rise rapidly. Above the critical scale, costs vary only slightly with size of firm. Moreover, in only a very few industries (the typewriter industry is perhaps the most striking example) does the critical scale represent a substantial percentage of the total market. These facts correspond well with beliefs about these matters that are widely held by businessmen.

We can say, then, that the characteristic cost curve for the firm shows virtually constant returns to scale for sizes above some critical minimum – s_m. Under these circumstances, the static analysis may predict the minimum size of firm in an industry with a known value of s_m, but it will not predict the size distribution of firms.

3. Stochastic models of firm size

In the context of a different theoretical framework, our limited knowledge of the shape of the long-run cost curve derived from static analysis might lead to much stronger predictions. This is, in fact, the case.

We postulate that size has no effect upon the expected

Table 7.1
Ingot capacities of ten leading steel producers (millions of net tons per year, based on capacity as of January 1, 1954).

| | Capacity | |
Producer	Actual[a]	Estimated
U.S. Steel	38.7	34.3
Bethlehem	18.5	17.1
Republic	10.3	11.3
Jones and Laughlin	6.2	8.5
National	6.0	6.8
Youngstown	5.5	5.2
Armco	4.9	4.8
Inland	4.7	4.2
Colorado Fuel and Iron	2.5	3.8
Wheeling	2.1	3.4
Total, 10 Companies	99.4	99.4

[a]*Source:* Actual from *Iron Age*, January 5, 1956, p. 289.

minimum scale for an efficient plant (Bain (1956, Appendix B)). There is much less basis for estimating, and much less consensus about, the minimum scale for an efficient firm.

Taking Bain's estimates of the minimum efficient plant size, on the one hand, and *Census of Manufacturers* data on the size distribution of plants, on the other, we have made some preliminary attempts to compare for several industries the minimum efficient scales suggested by these two sets of data. The results are listed in table 7.2. Our procedure was this: if there is a sharp increase in unit costs below some critical size, the number of plants in the industry below that size should be less than the number predicted from the Yule process. We plot cumulative numbers of plants against size on log paper, and look for sharp bends from a slope approximating −1 to a lower slope.

We have used census data for numbers of employees and converted these to percent of total value added by manufacture.

Table 7.2
Estimate of minimum feasible plant size.[a]

Industry	Bain estimate as percent of national market	Estimate from census data by Yule distribution as percent of total value added by manufacture
Flour and milling	0.05 to 0.25	0.07 to 0.19
Footwear	no minimum	0.03 to 0.07
Canned fruits and vegetables	no minimum	0.06 to 0.11
Cement	0.4 to 0.7	0.14 to 0.54
Distilled liquors (except brandy)	0.2 to 0.3	0.03 to 0.11
Petroleum refining	0.4 to 0.9	0.12 to 0.34
Meat packing	no minimum	0.3 to 0.7
Rubber tires and tubes	0.35 to 0.7	1.6 to 5.5
Rayon	1.0 to 3.0	0.14 to 0.37
Soap and glycerin	0.2 to 0.3	0.03 to 0.11
Cigarettes	1.0 or less	0.08 to 2.0
Fountain pens and mechanical pencils	1.3 to 2.5	0.06 to 0.16
Typewriters	5.0	5.7 to 14.1

[a]Minimum feasible plant size is that below which costs per unit rise substantially. The industries listed are those used by Bain, with seven omitted because of the inadequacy or incomparability of the data.

Sources: The Bain estimates were computed by multiplying his estimates of minimum efficient plant size (Bain (1956, table III, p. 72)) by the fraction of their size that was encountered before costs rose substantially (Bain (1956, Appendix B)).

The estimates from census data were computed by plotting the cumulative number of firms from the 1947 *Census of Manufacturers* against size in number of employees for the industries listed. Sharp breaks in the cumulative plot from a slope of -1 were taken as estimation points for the minimum feasible plant size and were converted to a percentage of total value added by manufacture from the same census tables.

Our measure is thus comparable to Bain's which is based upon the percentage that a plant represents of total national market.

The reader can draw his own conclusions as to how far the two estimating procedures lead to similar results. Since we have made no more than preliminary explorations, we do not wish to push the point too hard. It is clear, however, that the stochastic model provides some novel ways of interpreting the data on size distributions that may cast considerable light on the question of economies of scale. The argument runs as follows: if we take the stochastic model seriously, then any substantial deviation of the results from those predicted from the model is a reflection of some departure from the law of proportionate effect or from one of the other assumptions of the model. Having observed such a departure, we can then try to provide for it a reasonable economic interpretation.

In concluding this discussion of the data, we should like to emphasize a point made earlier – that the transition matrices may provide an even more valuable source of data about the process determining the sizes of firms and plants than the size distributions themselves. Since most of the empirical work to date has focused on the latter rather than the former, the reversal of emphasis initiated by the work of Champernowne, Hart and Prais, and others, is a very promising one.

5. Implications for economic policy

In discussions of the degree of competition in individual industries, various measures of degree of concentration have been used. Few of these have other than an empirical basis, and the values that are obtained depend, in ways that are only partly understood, on methods of classification, cut-off points, and the like. Among the frequently used measures are Lorenz's and Gini's coefficients of concentration.

As Aitchison and Brown (1957, pp. 111–16) argue, if we fit a distribution function to the observed data on the basis of a theoretical model, it is reasonable to base our measures of concentration on the parameters of the distribution function. Thus, they propose the standard deviation of the log-normal as an appropriate measure of dispersion, and show that the Lorenz and Gini coefficients can be expressed as functions of that statistic.

Similarly, if we use the Yule process to account for the distribution of firm sizes, our interpretation of the observed phenomena should be based on the estimated values of the parameters of the distribution. In the simplest case, the only one we have considered here, there is a single parameter, ρ. We have already provided an economic interpretation for this parameter in the previous section – it measures, in a certain sense, the rate of new entry into the industry. Hence, in this particular model, the concentration in an industry is not independently determined, but is a function of rate of new entry.

We may put the matter more generally. If firm sizes are determined by a stochastic process, then the appropriate way to think about public policy in this area is to consider the means by which the stochastic process can be altered, and the consequences of employing these means. As a very simple example, if the rate of entry into the industry can be increased, this will automatically reduce the degree of concentration, as measured by the usual indices. Similarly, if, through tax policies or other means, a situation of sharply increasing costs is created in an industry, this situation should cause a departure of the equilibrium distribution from the Yule distribution in the direction of lower concentration.

A third, and more complicated, example is this: the amount of "mixing" that takes place – reordering of the ranks of firms in an industry – depends on the dispersion of the columns of the transition matrix. The same equilibrium distribution may be produced with various degrees of mixing, since the latter can vary independently of the law of proportionate effect. Public policy might be concerned with the amount of mobility rather than with the resulting degree of concentration. As a matter of fact, a measure of mobility (for firms or individuals) would appear to provide a better index of what we mean by "equality of opportunity" than do the usual measures of concentration.

The net effect of approaching the subject of industrial concentration in this way will be to make the classical theory of the firm much less relevant to the subject, but theories of economic development and growth much more relevant. When we have a collection of adaptive organisms placed in a relatively stable environment, we can often make strong predictions about the resulting state of affairs by assuming that the system will come into a position of stable, adaptive equilibrium. When, however, the environment itself is changing at a rate that is large compared with the adaptive speeds of the organisms, we can never expect to observe the system in the neighborhood of equilibrium, and we must invoke some substitute for the static equilibrium if we wish to predict behavior. Our main objective in this chapter is to suggest the need for, and the availability of such a substitute with which to analyze the size distribution of firms.

6. Directions for research

We have emphasized the tentative character of our results, and should like to suggest in conclusion some directions of research that look exceedingly promising:

(1) We need to accumulate a body of knowledge about skew distribution functions and the processes that generate them that is comparable to the rich knowledge we possess about the normal, Poisson, exponential, and related distributions. We need to know more about the relations between the distributions and the generating processes, about efficient methods for estimating parameters, about the distributions of these estimates, and about efficient methods for choosing among alternative hypotheses.

(2) We need to develop stochastic models of economic growth that embody as much knowledge as we have, or can acquire, about the underlying processes.

(3) We need to re-examine the corpus of economic data to see

what part of it can profitably be explained or reinterpreted in terms of such economic models.

(4) We need to re-examine those principles of public policy that are based on static equilibrium analysis to see what part of them will remain and what part will be altered as stochastic processes begin to play a larger role in our explanation of economic phenomena.

In this chapter we have tried to suggest some of the directions in which inquiry may lead if it is guided by questions such as these.

CHAPTER 8

Business firm growth and size*

This chapter undertakes at the same time to increase the realism of the assumptions on which the stochastic models of firm size rest, and to improve the fit of the theory to the data. The simplest models, like those discussed in ch. 7, assume that the growth rates of individual firms are uncorrelated from one time period to another. Ch. 8 relaxes this assumption in a particular way. It assumes that the expected rate of growth of a firm depends not only on the total growth it has had in the past (i.e. its present size) but also on how recent that growth has been. Firms that have had recent rapid growth are postulated to have larger expected growth rates than firms of the same size whose growth took place earlier.

The new hypothesis might be viewed as a kind of "maturation" theory of organizations: they are born, at some time they may go through a period of rapid growth, at some later time that growth may slow down or even stop. The steady-state distribution for this model is treated analytically in ch. 11. In the present chapter, histories of economies behaving like the model are simulated, with the rate at which past growth is discounted as parameter.

The distributions generated in the simulation runs give good approximations to the Pareto distribution, the closeness of the approximation decreasing as the time discounting is made more rapid. The growth rates exhibit strong serial correlation, as would be expected, and the distributions are concave toward the

*We are indebted to the Ford Foundation for fellowship and research grants that made this work possible.

origin, hence providing a possible explanation for the concavity that has been noticed in nearly all of the empirical distributions.

It should perhaps be emphasized that it is not the serial correlation in this model that produces the concavity to the origin of the distributions, but rather the underlying "maturation" mechanism. In ch. 9 we will examine a somewhat different model of autocorrelated growth whose steady-state distribution is log normal.

1. Introduction

It is well known that highly skewed frequency distributions, similar to the observed distributions of sizes of business firms, can be generated by a number of related stochastic processes. All of the processes have at their core something like Gibrat's law of proportionate effect – the postulate that expected rate of growth is independent of present size. Several of these processes have been offered as explanations of the observed size distributions.

Stochastic explanations for the size distribution of firms have considerable interest for economic theory and policy. They interpret these distributions in terms of the dynamics of the growth process rather than in terms of static cost curves. If the assumptions on which they rest are correct, the models call for new statistical measures of the degree of concentration and new interpretations of the economic implications of concentration. These implications have been discussed in ch. 7. The present chapter is concerned with demonstrating that certain assumptions, known to be incompatible with the empirical data, that were used in the derivation of the earlier model (or in the description of that model in ch. 7) can be replaced by weaker and more realistic assumptions.

The adequacy of any theoretical explanation can be judged on two grounds: (a) the plausibility of its assumptions, their agreement with known facts; and (b) the goodness of fit of the derived distributions. Evaluating the goodness of fit of mathematical models like the ones under consideration here involves all of the theoretical difficulties of testing extreme hypotheses. Since these difficulties, although well known to mathematical statisticians, have received little discussion in the literature, we should like to pause to indicate their nature.[1] Suppose we wish to test Galileo's law of the inclined plane – that the distance, $s(t)$, traveled by a ball rolling down the plane increases with the square of the time: $s(t) = kt^2$, where k is a constant. We perform a large series of careful observations, obtaining a set of $[s, t]$ pairs from which we estimate k. The actual observations cluster closely around the fitted curve, but do not, of course, all fall exactly on it. To decide whether we have confirmed or refuted Galileo's law, we test whether the observed deviations of the observations from the fitted curve could have arisen by chance. Suppose the answer is that they could. Then we may conclude either (a) that Galileo's law is correct (i.e. at least not incompatible with these data) or (b) that our observations were not accurate enough to reveal its inadequacies. The more random "noise" there is in our data, in fact, the more likely that we will decide the law has been "confirmed."

Suppose, on the other hand, that the statistical test rejects the hypothesis. Then we may conclude either (a) that Galileo's law is substantially incorrect or (b) that it is substantially correct but only as a first approximation. We know, in fact, that Galileo's law *does* ignore variables that may be important under various circumstances: irregularities in the ball or the plane, rolling friction, air resistance, possible electrical or magnetic fields if the ball is metal, variations in the gravitational field – and so on, ad infinitum. The enormous progress that physics has made in three centuries may be partly attributed to its willingness to ignore for a time discrepancies from theories that are in some sense substantially correct.

Since no one has ever formalized the criteria for ignoring

---

[1] A succinct discussion with references will be found in Savage (1954, pp. 254–56).

discrepancies of this kind, and since the received body of statistical theory provides no suitable means for testing extreme hypotheses, we shall not test statistically the goodness of fit of the stochastic models. In the case of size distributions of firms, the observed distributions certainly "look like" Pareto, Yule, or log normal distributions, but there is no known satisfactory way to objectify the degree of resemblance. Since the observed distributions are radically different from those we would expect from explanations based on static cost curves (ch. 7), and since there appear to be no existing models other than the stochastic ones that make specific predictions of the shapes of the distributions, common sense will perhaps consent to what theory does not forbid – accepting the stochastic models as substantially sound.

Under these circumstances, however, our confidence in the proposed stochastic explanations of the size distributions may depend quite as much on how plausible we find the assumptions underlying the models as on our judgments of goodness of fit. If the assumptions are very strong, and particularly if they contradict known facts about business size and growth, we will be inclined to dismiss the stochastic explanations. If the assumptions are weak, and consistent with our empirical data, our confidence in the stochastic models will be correspondingly strengthened.

In §2 of this chapter we shall describe briefly the simplest stochastic models that have been used to derive the size distribution of firms and shall point out in what respects the assumptions incorporated in these models are contradicted by the empirical facts. In the following section we shall propose a model that removes what is perhaps the least acceptable assumption in the simple models: the assumption that the growth rates of individual firms in one period of time are uncorrelated with their growth rates in preceding periods. In the remainder of the chapter, we shall examine the distributions that are generated by the improved model.

2. Simple stochastic models

The simplest kind of stochastic process that will yield skew distributions like the observed ones is based on the following assumption:

Year-to-year changes in firm sizes are governed by a simple Markoff process in which the probabilities of the size changes of any specified percentage magnitudes are independent of a firm's present absolute size.

That is, each firm, under this assumption, has the same probability as any other firm of increasing or decreasing in size by 5 percent, 10 percent, or any other relative amount. This is Gibrat's law in its simplest and strongest form.

Let i be a measure of firm size – total assets, say – and $F(i)$ be the number of firms of size i or *larger*. (Thus, if there is only one firm of exactly size i, for all $i = 1, 2, \ldots, n$, $F(i)$ will be the *rank* of that firm in the industry.) Then, the process we have just described leads to an equilibrium size distribution which for large i is approximately

$$F(i) = \Gamma(\rho + 1)i^{-\rho}, \tag{1}$$

where ρ is a constant that depends on the rate at which new firms enter the industry, and Γ is the gamma function. (For details, see ch. 1 and ch. 4, eq. (24).) Function (1) is the Pareto distribution and is also, for large i, an approximation to the Yule distribution. Converting to logarithms, it yields the linear relation,

$$\log F(i) = -\rho \log i + a \text{ constant}. \tag{2}$$

The observed size distributions of business firms fit eq. (2) rather well (see ch. 7). However, distributions that are quite similar can be derived from processes based on weaker assumptions. First of all, it has been shown that we can replace the strong form of Gibrat's law, stated above, with the assumption that the expected percentage change in size of the *totality of firms in each size stratum* is independent of stratum. On this weaker assumption, the Yule, Pareto, or lognormal distributions considered in ch. 7 can still be derived. (See ch. 1 for more detailed discussion.) The stochastic model based on the weaker assumption is obviously also consistent (as the model based on the stronger assumption is not) with the following well-known facts:

(1) The transition matrices for changes in firm size from year to year show different relative (i.e. percentage) variance for the firms in different size groups – the relative variance decreasing with increase in size. Only the expected percentage changes for the totality of firms in a size group are independent of size (see Mansfield (1962)).

(2) The expected rates of change are certainly not equal, or nearly equal, for all individual firms. The simplest way to show this is to examine substrata for differences in expected rate of change – for example, substrata corresponding to different industry groups.

It is not clear, however, that the weaker assumption is consistent with a third fact:

(3) The expected rates of change for individual firms are not independent of changes in the prior years – there is serial correlation in the growth rates over at least short time periods.

The main purpose of the study reported here was to see whether a stochastic process based on some variant of Gibrat's law, but allowing for serial correlation from one period to another in the growth rates of individual firms, could also lead to distributions like those actually observed. If such a process could be found, it would open an additional direction for the generalization of stochastic explanations of firm size to fit a wider range of empirical findings. If it could not, grave doubt would be cast on the adequacy of stochastic models of this general kind for explaining the size of business firms.

We leave to later investigation some other important directions of generalization. In particular, the models described here do not admit mergers or decreases in size of individual firms. Our intuitions suggest that so long as the probabilities of merger or decline in size are roughly independent of stratum, they will not change much the equilibrium distributions, but this conjecture remains to be tested in future work.

3. A model with serial correlation

The implications of serial correlation in firm growth are not easy to trace analytically. Stochastic models admitting serial correlation have proved to be too complex to be solved explicitly in closed form for the equilibrium distributions. To cast some light on the question of whether a stochastic process with serial correlation could lead to distributions like those actually observed, we have carried out some Monte Carlo calculations with a class of process of this kind.

In the models to be discussed here, the identity of each individual firm is maintained from one time period to the next. The change in size of each firm is governed by a stochastic process, which depends on the size to which the firm has grown, but also upon the times at which its growth has taken place. For simplicity in our computations, we assume growth to take place in increments of unit magnitude. The probability that a firm will experience an increment in size during the next time period is assumed proportional to a weighted sum of the increments it has experienced in the past, where the weight of an increment decreases geometrically, at a rate γ, with the lapse of time since its occurrence.

We formalize these notions as follows: Let $y_i(k)$ be the change in size of the jth firm during the kth time interval, where $y_i(k)$ is either unity or zero (the firm either experiences a unit increment in size or remains the same size during any given time interval). Then the size of the jth firm at the end of the kth interval is simply

$$\sum_{\tau=1}^{k} y_j(\tau).$$ (3)

The expected increment in size of the jth firm during the $(k + 1)$st interval is

$$p[y_j(k + 1) = 1] = \frac{1}{W_k} \sum_{\tau=1}^{k} y_j(\tau)\gamma^{k-\tau}$$ (4)

where W_k is a function of time that is the same for all firms, and γ is the fraction that determines how rapidly the influence of past growth on new growth dies out.

Under these assumptions, large firms will, *ceteris paribus*, grow proportionately more rapidly than small firms – the *ceteris paribus* assumption being that the previous growth of the firms being compared took place at about the same times. On the other hand, firms that have experienced recent growth will grow more rapidly, *ceteris paribus*, than firms of the same size whose growth took place earlier.

We complete our model by adding assumptions about the total number of increments per time period, and the rate of entry of new firms. For simplicity in illustrating the behavior of models of this kind, we shall select the unit time interval so that there is exactly one increment per time period in the total assets of the entire industry. We shall assume that there is a constant probability, α, that this increment will be allocated to a new firm (which is therefore assumed to be one unit in size at the outset).[2] Under these assumptions, new firms will enter at an expected rate proportional to the rate of growth of the industry.

4. Details of the simulation procedure

Before we present some of the statistical results, we should like to describe in more detail exactly how the simulation of the

[2]Thus the size unit may be interpreted as the minimum efficient size of a firm in the industry (see Savage (1954, p. 608)); and a firm of, say, 15 units may be thought of as 15 times the minimum efficient size.

model just described was carried out. Readers who are not primarily interested in the procedure may prefer to skip the present section.

Suppose we wish to simulate the growth of an industry from its birth to an aggregated asset size of K units. (As suggested above, a "unit" may be interpreted as the total assets required for a firm of minimum efficient size in the industry.) We allocate the K units of assets one at a time; and for convenience, we shall describe the process as if the total size of the industry grows by one unit in each "time period." Stated otherwise, we take as the kth time unit ($k = 1, 2, \ldots, K$) the period during which the industry grows from ($k - 1$) to k in total assets. This kth unit of assets is assigned to some particular firm, either a new firm (which then achieves a size of one unit) or an existing firm (which then grows by one unit during the kth time period).

The allocation of the kth unit is made in two stages:

Stage I. This stage determines whether the unit is to be allocated to a new firm or an existing firm. We first draw a random number, a, from a rectangular distribution between 0 and 1. If $a \leq \alpha$, where α is a given constant, we create a new firm, the $n(k)$th firm, where $n(k)$ is the number of firms generated during the first k periods. We assign the kth unit of assets to this new firm, thus completing the allocation process for the kth time period without going through stage II. (Note that the total number of firms, $n(k)$, has been increased by one: $n(k) = n(k - 1) + 1$.) If, however, $a > \alpha$, then we assign the kth asset unit to one of the existing firms, as determined by stage II.

Stage II. For each firm in the industry, we keep track of two factors: the current *size* of the firm, and the current *growth potential* of the firm. By the size of the jth firm at the end of the ($k - 1$)th time period, $i_j(k - 1)$, we mean the total number of asset units that have been allocated to the jth firm up to that time. By the growth potential of the jth firm at the end of the ($k - 1$)th time period, $w_j(k - 1)$, we mean a weighted sum of the asset units that have been allocated to that firm up to that time.

The weights are assigned as follows: If we take as 1 the weight of the asset unit assigned during period $(k - 1)$, then the weight of the unit assigned during period $(k - \tau)$ will be $\gamma^{(\tau-1)}$, where γ is a proper fraction. That is, prior growth is assumed to create potential growth at a rate that falls off geometrically with the lapse of time since the prior growth occurred. We also keep track of the sum of the growth potentials for all firms: $W(k - 1) = \Sigma_j w_j(k - 1)$.

To assign the kth asset unit to a firm, we draw another random number b from a rectangular distribution between 0 and $W(k - 1)$, and assign the unit to the jth firm, where j is the smallest integer which satisfies

$$\sum_{l=1}^{j} w_l(k - 1) \geq b.^3$$

Then, the probability that the asset unit will be assigned to any particular one of the existing firms is proportional to the weight of that firm.

Thus, at the end of the Kth period, all K units of the assets of the industry will have been allocated among $n(K)$ firms. The distribution of assets is given by $i_j(K)$ $(j = 1, 2, \ldots, n(K))$. Then $F(i, K)$ is the number of firms whose size at the end of the Kth period is greater than or equal to i.

Since stage II of the process requires that there already be some firms in existence, the scheme is started off by providing a small initial population of firms. The initial conditions used in most of our simulations specified three already existing firms, one of 5 units, one of 3 units, and one of 1 unit. The final distribution is not entirely independent of the initial conditions, but tends to become independent as K grows large. (Except for the few largest firms, the distribution is very insensitive to the initial conditions even for a relatively small K.)

[3]For reasons of speed, the actual computation is a little different. Instead of recomputing each of the weights each time period, we simply increase the *new* weights geometrically with time. Thus the asset unit allocated during period k is assigned a permanent weight of γ^{-k}. Clearly, this leads to exactly the same probability of each firm being allocated the new asset unit, for all weights, under the modified scheme, are multiplied by the same factor, γ^{-k}.

5. A distribution generated by the model

As a first example of a distribution generated by this process, we consider a sample of 247 firms whose aggregate size, K, is 1 000 asset units, or an average of about 4 units per firm.[4]

The weights were reduced at a rate of 5 percent per time period [$\gamma = 0.95$]. Table 8.1 shows the frequency distribution of firms by size. The largest firm is 56 units. At the other end of the scale, there are 110 firms of one unit each. As is characteristic of these models, the distribution is highly skewed and the cumulative distribution is approximately linear on a double-log scale (fig. 8.1). Thus, the introduction of the weights did not change the general character of the equilibrium distribution.

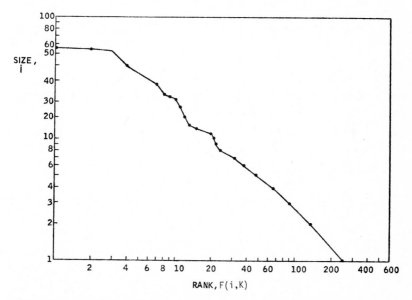

Fig. 8.1. Rank-size distribution of simulated "industry" of 247 firms.

[4]In generating this particular sample, α was not strictly constant but was set at $n(k)/k$. (The usual initial conditions, described in the previous section, were imposed to start the process off.) In generating the other samples reported here, α was fixed at a constant value. This change in assumption did not noticeably change the qualitative characteristics of the sample.

Table 8.1
Distribution of firms by size in a simulated "industry" of 247 firms.

Firm size (1)	Number of firms (2)	Weight w_j (3)	Cumulative weight Σw_j (4)	Cumulative percent of total assets (5)
56	1	–*	–	0.056
54	1	–	–	0.110
52	1	–	–	0.162
39	1	0.271	0.271	0.201
28	3	0.002	0.273	0.285
23	1	–	0.273	0.308
22	1	–	0.273	0.330
21	1	–	0.273	0.351
18	1	–	0.273	0.369
15	1	–	0.273	0.384
13	1	–	0.273	0.397
12	2	0.102	0.375	0.421
11	5	0.005	0.380	0.476
10	1	–	0.380	0.486
9	1	–	0.380	0.495
8	2	–	0.380	0.511
7	7	0.025	0.405	0.560
6	7	–	0.405	0.602
5	10	0.025	0.430	0.652
4	18	0.132	0.562	0.724
3	24	0.207	0.769	0.796
2	47	0.181	0.950	0.890
1	110	0.050	1.000	1.000
Total	247	1.000		

*Less than 0.001.

The final weights, after 1 000 time periods, were very far from proportional to the sizes of the individual firms. A mere 16 firms, out of the total of 247, accounted for almost all the potential for future growth. Each of these 16 firms had 1 percent or more of the total of weights. The remaining 231 firms had, *in the aggregate*, only about 1 percent of the total weight. Thus, at $k = 1\,000$, there was about one chance in four that the next increment would go to a new firm, since $\alpha = 0.247$; if it did not, there were 99 chances in 100 that it would go to one of the 16 firms with nonnegligible weights. These sixteen firms ranged over the whole size distribution: one was of size 39, one of size 12, one of size 7, one of size 5, two of size 4, four of size 3, and three each of sizes 2 and 1. In fact, these sixteen "growing" firms formed a skew distribution of almost the same shape as the entire distribution of 247 firms.

We see that Gibrat's law was very far from being satisfied for the individual firms. However, the growth potential was widely distributed over the size strata. In table 8.1 we also show (column 4) the aggregate of weights for all firms above a given size, and (column 5) the aggregate size of all firms above a given size. Thus, firms of size 28 and larger account for 27.3 percent of the total weight, but 28.5 percent of the total "assets." Their growth potential, in the aggregate, was therefore only slightly less than that postulated by Gibrat's law. The corresponding figures for firms of size 10 and larger are 38.0 percent and 48.6 percent, respectively; for firms of size 5, 43.0 percent and 65.2 percent, respectively; and for firms of size 3, 76.9 percent and 79.6 percent, respectively. Middle-sized firms, then, show less than average growth potential, firms of size 2 to 4, greater than average potential, firms of size 1, less than average.

6. The history of a simulated industry or economy

To get a better understanding of the growth patterns produced by the model, we made additional runs, recording the sizes of individual firms at successive time intervals. Data are presented in table 8.2 from a run of 1 000 time units, with α, the probability of assigning the next increment to a new firm, set at 0.2; and γ, the rate of discounting past growth, set at 0.95.

The table shows the sizes, at each 100 time intervals, of the 20 firms (out of a total of 209) that reached a final size of more than ten units. A number of interesting observations can be read from this table. Let us call the 20 firms whose histories are recorded in the table the "large" firms. First, older firms had not much greater chance of becoming large than younger firms. Eleven of

Table 8.2
Growth pattern of 20 largest firms in a simulated "industry" of 209 firms.

Firm number (in order of entry)	Period ending at time:									
	100	200	300	400	500	600	700	800	900	1 000
1	39	61	78	98	107	123	126	126	126	126
3	19	26	26	26	26	26	26	26	26	26
14	7	11	27	27	27	27	27	27	27	27
50				4	13	28	29	29	29	29
58				1	9	16	16	16	16	16
75					6	11	11	11	11	11
78					6	11	11	11	11	11
81					1	10	11	11	11	11
97					3	20	24	24	24	24
99					1	13	13	17	18	18
101						1	14	15	15	15
111						4	12	15	15	15
118							8	20	20	20
119							10	11	11	11
125							11	15	15	15
131							6	14	17	17
134							9	25	34	55
151								5	20	34
176									7	13
184									5	11
Total number of firms	24	43	61	81	103	118	141	166	189	209

the large firms entered the system in the first half of the time interval, and nine in the second half – an insignificant difference. There was some underrepresentation, among the large firms, of the very youngest; only three firms among the youngest 60 – i.e. the firms numbered 150 to 209 – had reached the large size.

If we consider the eight largest firms, those of size 20 and over, there is a little more relation between size and age. The very largest firm was the first to enter the system, and three other very large firms were among the first 50 to enter. On the other hand, the second and third largest firms are relative newcomers, having appeared during the last 40 percent and 30 percent of the system's history, i.e. after times 600 and 700, respectively.

We see, further, that firms which grow large experience most of their growth during the first 200 or 300 time periods after they enter the system, then reach a plateau. The very large size of firm 1 is associated with an abnormally long period of rapid growth (see table 8.3 and fig. 8.2). The firm grew at a virtually constant percentage rate until its growth stopped. Table 8.3 shows how, during its period of growth, its weight (probability of being chosen for the next increment) remained in the range 0.1 to 0.3. When the weight dropped, virtually to zero, during the 600th to 700th period, the growth stopped abruptly.

Table 8.3
Growth and growth potential of largest firm in
a simulated "industry".

Period ending	Firm size	Weight – w_j
100	39	0.273
200	61	0.306
300	78	0.223
400	98	0.143
500	107	0.118
600	123	0.257
700	126	0.007
800	126	–*
900	126	–*
1 000	126	–*

*Less than 0.001.

We do not wish to defend in detail the realism of the assumptions of the model or of the histories it produced. The time discount rate, γ, was almost certainly too high – underestimating the advantage of established firms – as compared with most real-life situations. What is interesting, however, is that a stochastic process that admits this very strong effect of recent growth in the determinants of future growth still produces the familiar kinds of skewed equilibrium distributions.

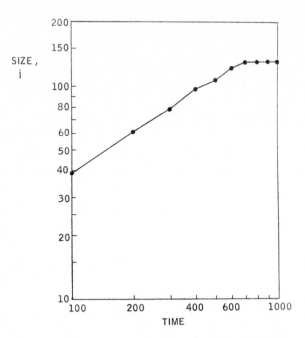

Fig. 8.2. Growth of largest firm in a simulated industry.

Clearly, Gibrat's law does not have to be assumed in any of its strong forms to produce this result.

7. Effects of parameters on the distribution

A substantial number of additional simulation runs, varying the parameters α and γ, provide information about the dependence of the equilibrium distribution upon these parameters. All of the cumulative distributions have a strong resemblance to that depicted in fig. 8.1. They are almost linear on a log scale. As α increases, the rank-order distributions (the distributions of $F(i, K)$, see fig. 8.1) acquire somewhat smaller initial slopes,[5]

---

[5]By initial slope we mean the angle of the curve to the horizontal axis at the bottom of the graph whose ordinate is $\log i$ and whose abscissa is $\log F(i)$.

and become slightly more concave to the origin. As γ decreases, the distributions again become considerably more concave to the origin, but acquire slightly *larger* initial slopes. The reciprocals of the slopes correspond to the parameter ρ in the equations (1) and (2). For the case where $\gamma = 1$, it can be shown that $\rho = 1/(1 - \alpha)$. The parameter ρ can be estimated, in the neighborhood of $i = 2$, by the equation

$$\rho_2 = -\frac{\log F(1, K) - \log F(2, K)}{\log 2}. \tag{5}$$

The values of ρ_2 for various combinations of α and γ are given in table 8.4. They range from 0.65, for $\alpha = 0.1$, $\gamma = 0.95$, to 1.4 for $\alpha = 0.4$, $\gamma = 1$.

Table 8.4
Estimates of ρ_2 for various values of α and γ.

α	γ			
	0.95	0.99	0.999	1
0.1	0.65	0.91	0.86	0.99
0.2	0.81	0.9	1	1.2
0.4	1.15	1.2	1.2	1.4

8. Conclusions

In this chapter we have described a new stochastic process, and have shown, by numerical simulation, that it generates equilibrium distributions that closely resemble the Yule distribution. In this process the expected growth rates of individual firms are assumed proportional to weights, where the weights are the time-discounted sums of previous increments in size. Thus the process incorporates a significant modification of the law of proportionate effect that allows substantial differences in the expected growth rates of individual firms.

CHAPTER 9

A model of business firm growth

This chapter, like the previous one, undertakes to account for serial correlation in firm growth rates. It does so, however, by dividing the growth of each firm into two components: its share of the industry growth, and an autocorrelated random (positive or negative) component peculiar to itself.

The random component of growth can be interpreted as resulting from more or less temporary advantages the firm is able to acquire over its competitors, in the form of an improved product, unusually able management, new marketing techniques, or the like. The degree of autocorrelation of the growth rates from one time period to the next depends, then, on how transient or permanent these advantages are on average – i.e. their time rate of discount. The model is fitted to data on the actual growth rates of large U.S. corporations to derive an estimate of this rate of discount.

The mathematical methods of this chapter are somewhat different from those of the other chapters. We consider the probability distribution at time t of the size of a firm that had a definite size at time zero. If the random components of growth are log normally distributed, the former probability distribution will also be log normal. To connect this analysis with the Yule and Pareto distributions discussed in the other chapters, we must consider a whole population of firms growing in this way, with new firms continually entering the population.

1. Introduction

A number of stochastic models, embodying various forms of Gibrat's law of proportionate effect, have been shown to generate skew distribution functions resembling the actual size distributions of business firms (see ch. 7). In ch. 8 we presented some results of the simultation of such a model permitting serial correlations over time in the size changes of individual firms. The aim of the present chapter is to carry further the analysis of autocorrelated growth, by proposing an economically meaningful scheme for its analysis, and applying the scheme to some data on large American firms.

In studying business firm growth, we often encounter cases where a firm suddenly acquires an impetus for growth. Perhaps by innovating in production or marketing processes, or perhaps as an effect of new management staffs or techniques, the firm grows much more rapidly than the other firms in the industry, as measured, say, by the ratio of the current firm size to its size in the previous time period. Thus, we may observe that, while most of the firms in the industry are growing at, say, 5 percent a year, some firms grow 10 percent.

Furthermore, a firm that grew 10 percent last year is likely to grow more rapidly than average again this year as a result of the carry-over effects of an innovation that occurred in a previous year on operations in subsequent periods. This carry-over becomes more and more likely as we shorten the length of the time period we are considering from a year to a month, week, or day. Moreover, on the average, a firm which grew rapidly in one year subsequently retains a greater share of the industry assets (or market share if sales are used as a measure of firm size) from that time on than do firms that have enjoyed only the average industry growth. Therefore, not only the growth rate over and above the average growth rate, but also the period when the extra growth took place are important factors in the individual firm's growth relative to the industry growth.

In this chapter, we develop a model to represent such characteristics of firms' growth, so that the process may be analysed further. In the final section we estimate the key parameter of the model for the recent growth of large American business firms.

2. The growth model

Let us represent by s_{jt} the size of the jth firm at the end of the tth period.[1] The size may be measured by the total assets of the firm or by its sales volume. We shall assume that there are N firms in the industry. For convenience, we shall consider a single industry, but will show later that the analysis may be applied to the economy of a given country as a whole.

Consider the relation that defines size ratios, G_{jt}:

$$s_{jt} = G_{jt}s_{j(t-1)}, \qquad t = 1, 2, \ldots, T. \tag{1}$$

Hence, the quantity G_{jt} may be called the growth ratio of the jth firm in the tth period. Let us decompose G_{jt} into two factors: one, a growth factor applicable to the jth firm only (the individual growth factor), g_{jt}, and the other, a growth factor that affects equally all firms in the industry (the industry growth factor), $\bar{g}_t$. Then we decompose G_{jt} into g_{jt} and $\bar{g}_t$ by the definitional equation:

$$G_{jt} = g_{jt}\bar{g}_t, \qquad t = 1, 2, \ldots, T. \tag{2}$$

Hence

$$s_{jt} = g_{jt}\bar{g}_t s_{j(t-1)}, \qquad t = 1, 2, \ldots, T. \tag{3}$$

Eqs. (2) and (3) are merely definitions of the growth rate factors. The industry growth ratio $\bar{g}_t$ affects the size of all firms in the industry equally and the individual growth factor, g_{jt}, is the residual of the jth firm's growth that has taken place in the tth period over and above the industry growth factor.

---

[1] In the prior chapters, the size of the firm has always been a discrete variable i, where $i = 1$ represents the minimum size s_m. In this and remaining chapters, we shall deal with a continuous and actual size variable and to recognize the difference we shall use a symbol s instead.

Eq. (2) defines only the product of the industry growth factor, $\bar{g}_t$ and the individual growth factor, g_{jt}; this product can be decomposed into its factors in any way that seems theoretically or statistically convenient. From the standpoint of the theoretical model, the individual growth factors should be defined so as to be statistically independent of the industry growth ratio. From a practical, statistical standpoint, however, it is satisfactory to identify the industry growth ratio with the quantity $\bar{g}_t = \sum_j s_{jt} / \sum_j s_{j(t-1)}$, that is, with the ratio of the size of the industry in the current period to its size in the previous period. Then g_{jt} is a measure of the change in the jth firm's share of market in the industry (using sales volume to measure size); so that if $g_{jt} = 1$, the ith firm has grown just rapidly enough to retain its share of market. With this definition, the statistical dependence of the average growth ratio on any individual growth factor will be too slight to bias significantly the estimates of parameters of the model, provided, of course, that the number of firms is relatively large.

From (3) we have

$$s_{jt} = \left(\prod_{\tau=1}^{t} g_{j\tau}\right)\left(\prod_{\tau=1}^{t} \bar{g}_\tau\right) s_{j0}. \tag{4}$$

That is,

$$\log s_{jt} = \sum_{\tau=1}^{t} \log g_{j\tau} + \sum_{\tau=1}^{t} \log \bar{g}_\tau + \log s_{j0}. \tag{5}$$

By means of definitions, we have attained, in eqs. (4) and (5), a decomposition of the size of the jth firm into a product of factors accounting for its growth. The first set of factors in the product reflects idiosyncratic events that distinguish this firm's history from the histories of other firms in the industry. The second set of factors determines the industry's growth. The final factor is the initial size of the firm.

Suppose that s_{j0} is given and $\bar{g}_\tau$, the industry growth ratio is also given for all $\tau = 1, 2, \ldots, t$. Then the only remaining factor in determining s_{jt} is $g_{j\tau}$ $(\tau = 1, 2, \ldots, t)$. We now assume that the quantities $g_{j\tau}$ satisfy the following hypothesis:

Hypothesis. The individual growth ratio g_{jt} of the jth firm in the tth period is the product of some power of the growth ratio $g_{j(t-1)}$ of the same firm in the $(t-1)$st period and a random factor ε_{jt}, which is distributed independently and identically for every firm and for every t, i.e.

$$g_{jt} = \varepsilon_{jt} g_{j(t-1)}^{\delta} \qquad (6)$$

where δ is a constant, and

$$g_{i1} = \varepsilon_{i1}. \qquad (7)$$

Notice that this hypothesis takes into account the following facts that we often observe in the analysis of firm growth.

(i) The expected value of the individual growth ratio is independent of the firm's size (Gibrat's law).

(ii) The individual growth ratio in one period is related to the individual growth ratio in the previous period (a single-period Markov process).

(iii) The individual growth ratio of a firm is determined independently from that of other firms. That is, factors that affect more than one firm are considered to be absorbed in the industry growth ratio $\bar{g}_t$.

(iv) With δ in the range $0 \leqslant \delta < 1$, an individual growth ratio in one period will have decaying effects on the ratios in subsequent periods. That is, a firm that grew more than (or less than) the industry growth rate in the previous period, namely $g_{j(t-1)} > 1$ (or $g_{j(t-1)} < 1$), on the average tends to grow more than (or less than) the industry growth rate in the current period but at a rate closer, on the average, to the industry growth rate than in the previous period.

Under our hypothesis we can develop the model as below. From (6) and (7), we have

$$\log g_{jt} = \log \varepsilon_{jt} + \delta \log g_{j(t-1)}$$
$$= \sum_{\tau=1}^{t} \delta^{(t-\tau)} \log \varepsilon_{j\tau}. \qquad (8)$$

Hence

$$\sum_{t=1}^{T} \log g_{jt} = \sum_{t=1}^{T} \sum_{\tau=1}^{t} \delta^{(t-\tau)} \log \varepsilon_{j\tau}$$

$$= (1 + \delta + \delta^2 + \cdots + \delta^{T-1}) \log \varepsilon_{j1} + (1 + \delta + \cdots$$
$$+ \delta^{T-2}) \log \varepsilon_{j2} + \cdots$$
$$+ (1 + \delta + \cdots + \delta^{T-k}) \log \varepsilon_{jk} + \cdots + \log \varepsilon_{jT}$$

$$= \sum_{t=1}^{T} \frac{1 - \delta^{(T-t+1)}}{1 - \delta} \log \varepsilon_{jt}. \tag{9}$$

Thus, from (5) and (9),

$$\log s_{jt} = \sum_{\tau=1}^{t} \frac{1 - \delta^{(t-\tau+1)}}{1 - \delta} \log \varepsilon_{j\tau} + \sum_{\tau=1}^{t} \log \bar{g}_{\tau} + \log s_{j0}. \tag{10}$$

We may remark that in the special case where $\log \varepsilon_j$ is normally distributed with mean zero and variance σ^2, ε_j has a log normal distribution. Let

$$x_{jt} = \sum_{\tau=1}^{t} \frac{1 - \delta^{(t-\tau+1)}}{1 - \delta} \log \varepsilon_{j\tau}. \tag{11}$$

Clearly, x_{jt} is normally distributed, since it is a weighted sum of independent random variables, each of which is normally distributed. In what follows, however, we assume independence of the $\log \varepsilon_{jt}$'s, but we do not assume normality. Since the mean of $\log \varepsilon_j$ is zero, the mean of x_{jt}, denoted by Ex_{jt}, is also zero, for

$$Ex_{jt} = E\left(\sum_{\tau=1}^{t} \frac{1 - \delta^{(t-\tau+1)}}{1 - \delta} \log \varepsilon_{j\tau} \right)$$

$$= \sum_{\tau=1}^{t} \frac{1 - \delta^{(t-\tau+1)}}{1 - \delta} E \log \varepsilon_{j\tau} = 0. \tag{12}$$

On the other hand, the variance of x_{jt}, denoted by Dx_{jt}, is given by

$$Dx_{jt} = D\left(\sum_{\tau=1}^{t} \frac{1 - \delta^{(t-\tau+1)}}{1 - \delta} \log \varepsilon_{j\tau} \right)$$

$$= \sum_{\tau=1}^{t} \left(\frac{1 - \delta^{(t-\tau+1)}}{1 - \delta} \right)^2 D \log \varepsilon_{j\tau}$$

$$= \frac{\sigma^2}{(1-\delta)^2} \sum_{\tau=1}^{t} [1 - 2\delta^{(t-\tau+1)} + \delta^{2(t-\tau+1)}]$$

$$= \frac{\sigma^2}{(1-\delta)^2} \left[t - 2\delta \frac{1-\delta^t}{1-\delta} + \delta^2 \frac{1-\delta^{2t}}{1-\delta^2} \right]. \tag{13}$$

Note that

$$\lim_{t \to \infty} Dx_{jt} = \infty, \tag{14}$$

and that for $0 \leqslant \delta < 1$,

$$Dx_{jt} \geqslant D\left(\sum_{\tau=1}^{t} \log \varepsilon_{j\tau} \right), \tag{15}$$

with the equality sign holding if and only if $\delta = 0$ or $t = 1$, since

$$\frac{1-\delta^{(t-\tau+1)}}{1-\delta} > 1 \quad \text{for all} \quad 1 \leqslant \tau < t, 0 < \delta < 1, t > 1. \tag{16}$$

Now let us return to eq. (10) for $\log s_{jt}$. If the second and third terms in the right-hand side of the equality are assumed to be determinate, the distribution function of $\log s_{jt}$ is completely determined by the distribution function of x_{jt}, except the position of the mean. Therefore, as t increases the probability density function for $\log s_{jt}$ becomes flatter and flatter, the distribution function approaching asymptotically:

$$F(x) = \tfrac{1}{2}. \tag{17}$$

On the other hand, we have

$$\begin{aligned} D(\log g_{jt}) &= D(\log \varepsilon_{jt} + \delta \log \varepsilon_{j(t-1)} + \cdots + \delta^{t-1} \log \varepsilon_{j1}) \\ &= D \log \varepsilon_{jt} + \delta^2 D \log \varepsilon_{j(t-1)} + \cdots \\ &\quad + \delta^{2t-2} D \log \varepsilon_{j1} \\ &= (1 + \delta^2 + \cdots + \delta^{2t-2})\sigma^2. \end{aligned} \tag{18}$$

Thus,

$$\lim_{t \to \infty} D(\log g_{jt}) = \frac{\sigma^2}{(1-\delta^2)}. \tag{19}$$

3. The multiplier

We can compare the limit of the variance of $\log g_{jt}$ just derived with the variance of the unweighted average, over time, of the ε's. Let us call the latter average v_{jt}, defined by:

$$v_{jt} = \frac{1}{t} \sum_{\tau=1}^{t} \log \varepsilon_{j\tau}. \tag{20}$$

Because of the independence of the ε_{jt}, we have immediately,

$$Dv_{jt} = \sigma^2. \tag{21}$$

Thus δ operates as a multiplier on the $\log \varepsilon_{jt}$, increasing the resulting variance in the growth ratios from σ^2 to $\sigma^2/(1 - \delta^2)$.

The empirical meaning of δ may be seen in the following manner. For simplicity, assume that

$$\log \bar{g}_\tau = a \quad \text{for all} \quad \tau = 1, 2, \ldots, t - 1, t, t + 1, \ldots, \tag{22}$$

where a is a constant and

$$\log \varepsilon_{j\tau} = 0 \quad \text{for all} \quad \tau = 1, 2, \ldots, t - 1. \tag{23}$$

Then $\log s_{j\tau}$ is given by

$$\log s_{j\tau} = a\tau + \log s_{j0} \quad \text{for} \quad \tau = 1, 2, \ldots, t - 1. \tag{24}$$

Suppose that $\log \varepsilon_{jt} \neq 0$, while $\varepsilon_{j\tau} = 0$ for $\tau = 1, \ldots, t - 1$, and that δ is equal to zero. Then the effect of $\log g_{jt}$ is to make a parallel shift of the time path for $\log s_{j\tau}$ by the quantity $\log \varepsilon_{jt}$ (see fig. 9.1).

When $\delta = 0$, there is no carry-over effect on the growth in the subsequent periods, hence the line after the shift will be parallel to the original one if the subsequent terms $\log \varepsilon_{j\tau}$, for $\tau > t$, are all zero.

Next, consider the case where $0 < \delta < 1$ and $\log \varepsilon_{j\tau} = 0$ for all τ except $\tau = t$. Then the time path of $\log s_{jt}$ is shifted by the quantity $\log \varepsilon_{jt}$ at the period t. It is shifted again in the period $t + 1$ by the quantity $\delta \log \varepsilon_{jt}$; in the period $t + 2$ by the quantity $\delta^2 \log \varepsilon_{jt}$, and so on. The actual growth curve, then, approaches asymptotically (fig. 9.2),

$$\log s_{j\tau} = a\tau + \log s_{j0} + \frac{1}{1-\delta} \log \varepsilon_{jt}. \tag{25}$$

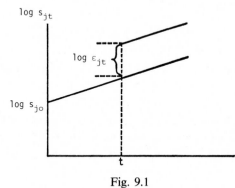

Fig. 9.1

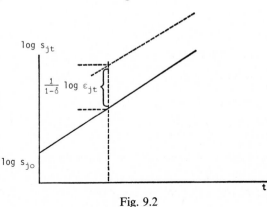

Fig. 9.2

4. A two-level generalization

The stochastic model is readily generalized to admit more than two causes of change in size. For example, the growth ratio for each firm might be expressed as the product of three factors: a ratio for the economy as a whole, a factor describing the growth of the firm's industry relative to the economy as a whole, and a factor expressing the growth of the individual firm relative to its industry. Then eq. (2) would be replaced by

$$G_{jlt} = g_{jlt}g_{lt}\bar{g}_t, \qquad t = 1, 2, \ldots, T, \tag{2'}$$

where $\bar{g}_t$ is the average rate for the economy, g_{lt} the growth factor associated with the lth industry, and g_{jlt} the factor associated with the jth firm in the lth industry.

Alternatively, we can combine the first two factors in (2'), arriving again at a product of two factors, the first of which reflects the joint effect of the idiosyncratic growth of the firm's industry relative to the economy and of the individual firm relative to the industry. In the next section we shall use this latter interpretation, formally identical with the original model of equation (2), to analyse some growth rates in the American economy.

5. Growth of large American firms

To illustrate the application of the model, we have estimated δ for the recent growth of large American firms. The data are the sales of the ninety six largest American firms, obtained from the *Fortune* tabulation,[2] for the years 1954, 1958, and 1962. A four year time interval was used so that the middle-run growth trends of individual firms would not be swamped by short-run business cycle fluctuations. Defining, as above, $\bar{g}_t = \Sigma_j s_{jt}/\Sigma_j s_{j(t-1)}$, the economy growth ratio was found to be 1.27 both for the four year period 1954–1958 and for the four year period 1958–1962. (This corresponds to a growth rate of about six percent per annum.) These quantities, inserted in eq. (3), provided estimates for the g_{jt} for the same two time intervals (call them g_{j1} and g_{j2}, respectively). Inserting the logarithms of these growth factors in

---

[2]Data for four of the one hundred largest firms in 1962 were not usable, because the data were not available for all three years, or because large scale mergers had made the data entirely noncomparable. The smaller noncomparabilities from year to year that undoubtedly exist for some of the remaining firms were simply ignored.

eq. (8), the method of least squares was used to estimate δ. The regression equation is

$$\log g_{j2} = 0.35 \log g_{j1} - 0.00034, \quad \text{or } \delta = 0.35. \quad (26)$$

Thus δ, the factor measuring the degree of persistence of sudden growth, was slightly greater than one third for large American firms over a four year period. A firm that experienced an unusually rapid growth in the first four year period could expect a greater than average growth in the second four year period. But the logarithm of the ratio measuring the excess would be, on the average, only one third as large during the second period as during the first. Thus, a firm that doubled its share of market (i.e. of the total economy) in the first four years ($g_{j1} = 2$), could be expected, on the average, to increase its share of market by about twenty eight percent in the second four year period (for $\log (1.28) \sim 0.105 = 0.35 \,(\log 2)$). Rapidly growing firms "regress" relatively rapidly to the average growth rate of the economy.

The same point may also be stated using eq. (25). Since for these data, $1/(1 - \delta) = 1.54$, a firm that experienced a "windfall" growth of magnitude $\log \varepsilon_{jt}$ during the first four year period could expect a total effect of this "windfall" upon $\log s_{jt}$ of 1.54 $\log \varepsilon_{jt}$ before its growth rate returned again to the average for the economy.

Since our data provide only two time intervals for comparison, they do not allow us to test the assumption that the $\log \varepsilon_{jt}$ are distributed independently for all time periods.

6. Conclusions

In this chapter we have proposed a model of business firm growth that decomposes the growth of a firm into an industry-wide component and a component peculiar to that firm. We have developed a Markov-process model for the individual component of growth, and have shown how to estimate the key parameter of the model, a parameter measuring the persistence of spurts in growth.

CHAPTER 10

Effects of mergers and acquisitions on concentration

A second limitation of the simplest stochastic models of business firm size, in addition to the fact that they do not allow for serial correlation in growth, is that they ignore the occurrence of mergers, acquisitions, and dissolutions. This chapter and the next one examine the effect of mergers and acquisitions upon the size distribution.

A question of particular interest, from the standpoint both of theory and of antitrust policy, is the effect that mergers will have upon the degree of industrial concentration measured by the slope of the distribution on the double-log scale. The chapter derives some conditions under which mergers will neither increase nor decrease the concentration, and presents some U.S. data for 1956–57 that appear to satisfy these conditions. (It should be pointed out that the more extensive data examined in ch. 11 do not fully support these conclusions as applied to the whole period from 1948 to 1969.)

The analysis of ch. 10 provides an answer to an apparent paradox that had been remarked upon in the literature: that there was little, if any, change in the slope of the Pareto curve for American industry during the first half of this century, even though there had been several periods of vigorous merger and acquisition activity.

1. Introduction

It is the central purpose of this chapter to explain a paradox: on the one hand, mergers and acquisitions of industrial firms have been exceedingly numerous in the U.S. economy during the past sixty or seventy years, and during the last several years they have reached epidemic proportions. On the other hand, the most careful studies of the degree of industrial concentration in U.S. industry over this same period show that overall concentration, as measured by the slope of the Pareto curve, has remained substantially constant – perhaps has even decreased, but certainly has not increased to an important extent. We propose to show that these two sets of facts – a high rate of mergers and acquisitions, on the one hand, and constancy of a concentration measure, on the other – far from being paradoxical or contradictory, follow from quite plausible (and empirically supported) assumptions about the merger and acquisition process.

Since the fact that the actual level of concentration has tended to remain approximately constant is contrary to popular folklore, we wish to call attention to the careful studies of this matter by M. A. Adelman, who reported his findings at the 1964 hearings of the U.S. Senate Subcommittee on Antitrust and Monopoly. Adelman concludes that the concentration ratio averaged over industries "appears to have declined substantially from 1901 to 1947, and not to have changed much since then" (U.S. Senate (1964, p. 231)). With respect to a more recent period, he says, "I do not see any possible escape from the conclusion that 'overall concentration' in the largest manufacturing firms has remained quite stable over a period of 30 years, from 1931 to 1960. I cannot conceive of any circumstances which could so affect the statistics that they failed to register an increasing concentration, taking place over so long a period of time" (U.S. Senate (1964, p. 237)). Finally, he laments the lack of "a logically consistent theory explaining why concentration should be stable. If we had the theory, then the actual past constancy would be a verification, and we would have some reason to think that if the conditions remained the same, the result should too" (U.S. Senate (1964, p. 240)).

It is our aim to provide the logically consistent theory that Adelman calls for. To do so, we must begin with the known facts about the distribution of firms by size.

It has been known for many years that the relation between the size s of a firm (measured by sales or gross assets) and the rank order r of the firm by its size in a population (for example, an industry, a nation, etc.) can be expressed approximately by the equation

$$sr^\beta = M, \tag{1}$$

where β and M are constants (Pareto law).[1] Taking logarithms (to any base),

$$\log s = \log M - \beta \log r, \tag{2}$$

and we see that $\log s$ and $\log r$ are linearly related as shown in fig. 10.1, where $\log M$ is the intercept at the vertical axis and $\beta = \tan \theta$.

From (1) we see that the size ratio of two firms s_1/s_2 is equal to the reciprocal of $(r_1/r_2)^\beta$, that is, the rank ratio r_1/r_2 raised to the power β. Thus, as the rank is doubled from r to $2r$, the size is decreased from s to $s/2^\beta$. In such a distribution, the largest firm sells 2^β times as much as does the second largest firm, the latter selling 2^β times as much as does the fourth largest firm, and so on (if sales are used as the size measure). Therefore, the larger the β, the greater the difference in size between two firms with a given ratio of their ranks.

This constant β, then, may be used as a measure of the degree to which business is concentrated in the larger firms in an industry or an economy. In contrast to other types of concentra-

---

[1] Compare this with the discrete Pareto distribution given in (42) in ch. 3, namely $F(i) = i^{-\rho}$. In (1) above, i is replaced by s, which is a continuous variable; $F(i)$ by r/n, where n is the number of firms in the population; and ρ by $1/\beta$. Hence, (42) in ch. 3 is equivalent to $r/n = s^{-1/\beta}$ or $sr^\beta = n^\beta$. By definition, M, which is equal to n^β, is the size of the largest firm.

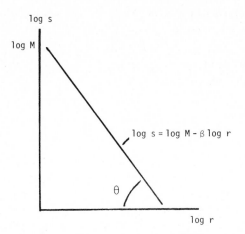

Fig. 10.1. Pareto distribution of firms (r = rank of the firm; s = size of the firm; $\beta = \tan \theta$, the concentration measure).

tion measures, which require an arbitrary cutoff point (for example, the market share of the 5, 25, 50, or 100 largest firms, etc.), the concentration measure β takes advantage of the linearity of the relationship between log s and log r to avoid setting such a cutoff point. The ch. 8 study of overall distributions, using data from 1956 *Fortune* directory and Hart and Prais (1956), indicates that for British firms $\beta = 0.474$ and for U.S. firms $\beta = 0.448$.

In this chapter, we want to analyze the effects of mergers and acquisitions on the overall firm size distribution, in particular their effects on the concentration measure β. For this purpose, we shall develop in the next section a mathematical model of the effects of mergers and acquisitions on the concentration measure β and then, in §3, analyze empirical data on large firms in the United States to see what they show about the actual effect of mergers and acquisitions.[2]

In §2 we shall be particularly interested in determining conditions under which the concentration measure will be unaffected

---

[2]A more complete model would also take account of spin-off and divestiture, but we shall not undertake this extension of the theory here.

by mergers and acquisitions, while in §3 we shall show that these conditions are reasonably satisfied by data on mergers and acquisitions in the United States in 1956 and 1957 – thus providing an explanation for the observed stability over time of the concentration measure.

2. A model

We can calculate the effects of mergers and acquisitions on the concentration measure as though the merger process occurred in two stages. In the first stage, we determine the size distribution after removal of the firms that disappear or are acquired (we shall call them "acquired firms") through mergers and acquisitions in a given period. Then in the second stage, we determine the size distribution after the market shares or the assets of the acquired firms have been distributed among the surviving firms.[3] Total sales or assets of all the firms are conserved over the two stages, taken together.

2.1. The disappearance process

In order to proceed with the first stage, let us compare the firm size distribution of all firms in the population with the firm size distribution of surviving firms, that is, firms other than acquired firms. We make the assumption that both distributions are of the form given by (1), involving a linear relation between $\log s$ and $\log r$. If we let r be the premerger rank of a surviving firm with size s and let r' be the postmerger rank of the firm, we have:

$$\log s = \log M - \beta \log r, \tag{3}$$

[3]Ideally, we would like to obtain the correlated data of the size of acquiring firms distributed by the size of acquired firms and analyze their relation. However, such data are very difficult to obtain (see, however, Federal Trade Commission (1955)). Furthermore, since we are interested in the effects of mergers and acquisitions on the concentration measure, rather than in the growth of individual firms, the above two-stage analysis is adequate for our purposes.

and

$$\log s = \log M' - \beta' \log r', \tag{4}$$

where M' and β' are the parameters for the postmerger distribution. Note that s, the size, is unchanged, since the assets of the acquired firms have not been distributed in this first stage. From these two equations, we can express r' in terms of r:

$$\beta' \log r' = \log \frac{M'}{M} + \beta \log r, \tag{5}$$

from which we get:

$$r' = \left(\frac{M'}{M}\right)^{1/\beta'} r^{\beta/\beta'}. \tag{6}$$

The fraction of the firms with rank r or smaller that survives is

$$\frac{r'}{r} = \left(\frac{M'}{M}\right)^{1/\beta'} r^{(\beta/\beta')-1}, \tag{7}$$

and the marginal survival probability is

$$\frac{dr'}{dr} = \frac{\beta}{\beta'} \left(\frac{M'}{M}\right)^{1/\beta'} r^{(\beta/\beta')-1}. \tag{8}$$

This is the survival probability of a firm with initial rank r and size $s = Mr^{-\beta}$.

Note that $\beta' = \beta$, that is, no change occurs in the concentration measure if and only if the survival probability is the same for all firms. For in this case, dr'/dr becomes independent of r. $[r^{(\beta/\beta')-1} = r^0 = 1.]$ In general, the ratio of the survival probabilities of two firms whose rank ratio is r_1/r_2 is given by $(r_1/r_2)^{(\beta/\beta')-1}$. For each doubling of rank, the survival probability is reduced by the factor $2^{(\beta/\beta')-1}$. Thus, the concentration measure after the disappearance of acquired firms would be doubled if the survival probability of a firm of rank $2r$ were only $2^{1/2-1} = 0.7$ times as large as the survival probability of a firm of rank r. Similarly, in order to cut in half the concentration measure after the disappearance of acquired firms, the survival probability of a firm of rank $2r$ must be $2^{2-1} = 2$ times as large as the survival probability of a firm of rank r.

2.2. The allocation process

Having analyzed the disappearance process, let us move on to the second stage and analyze the allocation process, the process of allocating the market share or the assets of acquired firms among the surviving firms. If we let s be the preallocation size of a firm whose postmerger rank is r' and s'' be the postallocation size of the firms, we have

$$\log s = \log M' - \beta' \log r' \tag{4}$$

and

$$\log s'' = \log M'' - \beta'' \log r', \tag{9}$$

where M'' and β'' are parameters of the postallocation distribution. From these two equations, we can express s'' in terms of s:

$$\log s'' = \log M'' - \frac{\beta''}{\beta'} \log M' + \frac{\beta''}{\beta'} \log s, \tag{10}$$

from which we get

$$s'' = M''(M'^{-\beta''/\beta'})s^{\beta''/\beta'}. \tag{11}$$

Thus, the ratio of new to initial size, due to the allocation, is

$$\frac{s''}{s} = M''(M'^{-\beta''/\beta'})[s^{(\beta''/\beta')-1}]. \tag{12}$$

Note that $\beta'' = \beta'$, that is, there is no change in the concentration measure if and only if each surviving firm increases its size, as a result of allocation, by a constant percentage of its preallocation size. The ratio of the percentage growth of two surviving firms whose preallocation sizes are s_1 and s_2 is given by $(s_1/s_2)^{(\beta''/\beta')-1}$. If $s_1 = 2s_2$, the percentage growth of the larger firm is $2^{(\beta''/\beta')-1}$ times that of the smaller firm. Thus, in order to double the concentration measure, after allocation, the percentage growth of a size $2s$ firm must be $2^{2-1} = 2$ times as large as the percentage growth of a size s firm. Similarly, in order for the concentration measure to decrease by one-half, the percentage of a size $2s$ firm must be $2^{1/2-1} = 0.7$ times that of a size s firm.

3. Empirical data

Empirical data for estimating the effects of mergers and acquisitions on the concentration measure are scarce. However, we were able to make crude estimates of the effects in the following manner.[4]

3.1. The disappearance process

First, we selected from the mergers and acquisitions reported by the Select Committee on Small Business (U.S. House of Representatives (1962)) those that took place in 1956 and 1957. The sales volumes of acquired firms in 1956 and 1957 were obtained from *Moody's Industrial Manuals* (1956, 1957), and they were grouped by sales volume as shown in table 10.1. The frequencies

Table 10.1
Survival probabilities of large American firms, 1956–57, by sales volume groups.

Sales volume	Total number of firms in *Fortune* 500 (A)	Number of firms acquired in 1956–57 (B)	Survival probability (C) (C = [A − B]/A)	Disappearance rate (D) (D = B/A)
Above $500 million	73	2	0.973	0.027
$500–$200 million	108	6	0.944	0.056
$200–$100 million	126	4	0.968	0.032
$100–$55 million	193	7	0.964	0.036

are compared with the distribution of 1955 sales volumes of firms reported in the *Fortune* 1956 directory prior to the mergers and acquisitions in question.

The fact that the survival probabilities (C) are all close to each other and show no apparent trend seems to indicate that β' is not much different from β.[5] There are substantial fluctuations in the

[4]Arvind Jain helped us in preparing the data.
[5]$\chi^2 = 1.24$, where $\chi^2_{0.25}(3 \text{ df}) = 1.21$ and $\chi^2_{0.50}(3 \text{ df}) = 2.4$.

disappearance rates (D), but the absolute numbers are small (B) and the fluctuations lie within sampling error and are not systematically related to firm size.

3.2. The allocation process

In the second stage, we took the same data of the Select Committee on Small Business (U.S. House of Representatives (1962)) for mergers and acquisitions that took place in 1956 and 1957. They are classified by the sales volumes in 1960 of the acquiring companies. Therefore, we calculated the total sales volume of all firms that were acquired by firms whose sales volume in 1960 fell in a specified size group (see table 10.2). This

Table 10.2
Ratio of sales volume of acquired firms to total sales of firms classified in sales volume groups.

Sales volume	Sales volume of firms in *Fortune* 500 (A)	Total sales volume of acquired firms, acquired by firms in the specified sales group (B)	B/A
Above $500 million	$130 077 million	$4 305 million	0.0331
$500–$200 million	43 411 million	582 million	0.0157
$200–$100 million	21 513 million	808 million	0.0376
$100–$72 million	9 723 million	157 million	0.0161

total for acquired firms was then compared with the total sales volume of all firms in the corresponding size group (based on the *Fortune* list for 1960). It is not apparent from these data that there is any substantial correlation between the size and growth rate by allocation. On the whole, the growth rate is relatively independent of size.

Obviously, we need more data to make a definitive statement about the effect of mergers and acquisitions on the concentration measure. However, our analysis does suggest tentatively that mergers and acquisitions do not greatly affect the Pareto curve slope. We shall consider some implications of this hypothesis from the overall viewpoint of the firm size distribution in the next section.

4. Gibrat's law for mergers and acquisitions

Intuitively, it appears to be unreasonable that mergers and acquisitions should not affect the concentration measure. However, that they need not can be well illustrated by the following example. Suppose that each firm in a population whose size distribution is given by (1) merges with another firm of equal size.[6] A firm of size s with rank r before merger now becomes a firm of size $s'' = 2s$ with rank $r'' = r/2$. The new distribution after mergers is, then,

$$\log s'' = \log M'' - \beta'' \log r''.$$

By substituting $s'' = 2s$ and $r'' = r/2$, we obtain

$$\log 2s = \log M'' - \beta'' \log \frac{r}{2},$$

or

$$\log s = \log (M''2^{\beta''-1}) - \beta'' \log r. \tag{13}$$

By comparing this with (2), we see that

$$M'' = 2^{1-\beta''}M, \tag{14}$$

and

$$\beta'' = \beta. \tag{15}$$

Thus, the concentration measure is unaffected, although the intercept at the log s axis (the size of the largest firm) is changed from M to $2^{1-\beta''}M$. Therefore, the size ratio of a large firm (rank r_1) and a small firm (rank $r_2 > r_1$) remains at $(r_1/r_2)^\beta$, unaffected by

---

[6] Although the distribution given in (1) implies that the size of each firm is different from every other firm, let us here assume that two firms with "nearly" equal size merge.

mergers.[7] This size ratio for a given rank ratio is one of the most important aspects of business concentration, since it expresses a relative strength of a larger firm over a smaller firm. The concentration measure β thus expresses this important aspect. It is with respect to this concentration measure that we want to see the effects of mergers and acquisitions on the firm size distribution. Of course, our results do not imply constancy of different concentration measures that might be used to measure other aspects of concentration, for example, share of assets held by the N largest firms.

If we plot the firm size distribution of the 500 largest firms in the *Fortune* list over the last twenty years or so, we see that the concentration as measured by the shape of the Pareto curve is relatively unchanged.[8] This supports Gibrat's law – that the growth rate is independent of size. The annual growth of the firms takes the form of a parallel upward shift in the size distribution, the degree of shift depending on the growth rate that is applicable to all firms regardless of their size. However, this observation supports Gibrat's law for the overall growth of firms but not necessarily for the growth by mergers and acquisitions alone. The overall growth of firms consists of internal growth (due to mergers and acquisitions) and external growth (due to growth from sources outside the population). That overall growth satisfies Gibrat's law does not necessarily mean that internal growth and external growth each satisfy Gibrat's law individually, since deviations from the law may cancel with each other to produce an overall Gibrat's effect. However, our data and analysis support the proposition that the internal growth alone does follow Gibrat's law. This implies that the

[7] The total number of firms in the population appears to be cut in half. However, if the population is defined by using a minimum size as a cutoff point, and if the mergers-with-equals are assumed also to take place among the subminimal firms, the total number of firms in the population after mergers is greater than half of the total number of firms in the population before mergers, due to new firms attaining above-minimum size by merger.

[8] The actual distributions have some curvature. See the discussions in chs. 8 and 9, which explain how such curvatures may result from assumptions on the growth of firms.

external growth also follows Gibrat's law, since, if a parallel overall shift in the distribution consists of two parts, one of which is a parallel shift, the remaining part must also be a parallel shift.

CHAPTER 11

Departures from the Pareto distribution*

This final chapter returns to, and carries further, several issues that have been raised in previous chapters: autocorrelation in growth rates, the effects of mergers and acquisitions upon concentration, and the concavity to the origin of the observed distributions.

First, an analytic solution is obtained for the autocorrelated growth model presented in ch. 8, where the process was analyzed by simulation. The analytical characterization of the model greatly improves our understanding of the autocorrelated growth process.

Second, the effect of mergers and acquisitions upon the firm size distribution is re-analyzed with the help of much more extensive and detailed data than those used in ch. 10. Using Federal Trade Commission data, the actual firm size distribution in 1969 is compared with the hypothetical distribution that would have been observed if all mergers were "undone." The results show that, for the time period in question, mergers and acquisitions had a detectable effect upon the concentration measure, as well as upon the concavity to the origin of the firm size distribution.

1. Concavity of the firm-size distributions

It has been observed repeatedly that the size, s, of a firm and its rank, r, in an industry or in an economy satisfy approximately

*This project has been supported in part by a grant from the National Science Foundation.

the Pareto law (see, for example (Steindl, 1965)),

$$sr^\beta = M, \tag{1}$$

where β and M are constants. The size may be measured by annual sales, total assets, or the number of employees. The rank is based on the selected size measure, the largest firm being assigned rank 1.

In a previous chapter we used the constant β as an indicator of the degree of concentration in the population because it indicates the frequencies of large firms relative to smaller firms. For example, if the rank of a firm X is twice the rank of a firm Y, the size of X is $2^{-\beta}$ times the size of Y, since $s_X = Mr_X^{-\beta} = Mr_Y^{-\beta}2^{-\beta} = 2^{-\beta}s_Y$. The larger the β, the greater the relative size of a large firm (small rank) compared with a smaller firm (large rank). Clearly, M is the size of the largest firm.

When $\log s$ is plotted against $\log r$, the relation is a straight line:

$$\log s = \log M - \beta \log r. \tag{2}$$

Thus, β is the slope of the line on a double-log scale.

The empirical data on firm sizes fit this distribution reasonably well as a first approximation (Steindl (1965)). However, we often encounter firm-size distributions for the economy that have some concave (downward) curvature. For example, the *Fortune* data for 1969 (*Fortune* (1970a, 1970b)) show a considerable curvature for the size distribution based on amount of assets (total assets after accumulated depreciation) for the 831 largest industrial firms. This distribution is shown in fig. 11.1.[1] The straight line through the empirical data is the Pareto curve fitted to the tenth largest firm and the eight hundred thirty-first firm.[2]

[1] Empirical rank-size distributions based on sales size and on assets size are almost indistinguishable except for the difference in the intercept as can be seen by plotting data from *Fortune* (1970a, 1970b).

[2] It has been shown in ch. 1 that the Pareto curve can be derived from Gibrat's law, which states that the percentage growth rate of a firm is distributed independently of its size.

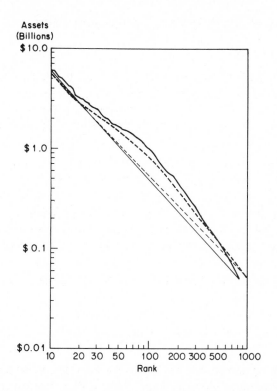

Fig. 11.1. Solid curve: actual rank-size relationship as of December 31, 1969. Broken curve: estimated rank-size relationship as of December 31, 1969, when the effects of mergers and acquisitions in 1948–69 were eliminated. Solid line: theoretical rank-size relationship (the Pareto curve) for actual data (connecting the sizes of the tenth and eight hundred thirty-first firms). Broken line: same as solid line for estimated "mergerless" data.

(The first nine firms are ignored in fitting the curve in order to take advantage of the statistical averaging incorporated in the cumulative distribution function except at its extreme tail.)

As shown in fig. 11.1, the actual size of a firm around rank 100 is almost twice as large as the "theoretical" size predicted from the straight line. That is, actual assets of the one hundredth firm are $1.01 billion while "theoretical" assets are $0.50 billion. Of course, many lines can be drawn on the figure as "theoretical" distributions. The above illustration is intended only to show the significance of the curvature of the actual firm-size distributions.[3]

The ratio of the actual size to the theoretical size of a firm with a given rank may be called a "size variance." Then, the downward concavity of the distribution shows an upward size variance for middle-rank firms relative to small- or large-rank firms, where the size variance is measured from the Pareto curve.

Faced with this rather significant departure from the theoretical distribution, one might abandon the Pareto curve and look for other distributions which, when the cumulative frequencies are plotted on a double-log scale, show curvature similar to the empirical distributions.[4] However, the Pareto curve is derivable from simple and economically plausible assumptions as shown in ch. 1, namely, size independence of percentage growth rate (Gibrat's law) and constancy of the entry rate. Both assumptions have significant economic implications. Hence, it seems to be profitable to attempt to retain the Pareto curve as a first approximation and to explain the deviations from this distribution by considering additional economic factors that violate the assumptions of the derivation, and hence could produce perturbations.

The purpose of this chapter is, therefore, to explore the ways of explaining the size variance by economic processes that can be observed empirically. We consider two alternative explanations. The first extends ch. 8, which analyzed by simulation the effects of autocorrelation of growth, giving a firm with a history of recent growth a better chance for further growth than a firm of the same size whose growth had taken place in the distant past.

Under Gibrat's law, a unit increase in the size of a firm

[3]The selection of the tenth and the eight hundred thirty-first firms is arbitrary. However, we use the "theoretical" distribution only for purposes of illustration. Our conclusions, especially with regard to the effect of mergers and acquisitions on the size distribution, would hold equally well for other methods of fitting a Pareto curve to the data. We have therefore used the simplest possible method of fitting – drawing the curve through two points in the cumulative distribution.

[4]The lognormal and negative binomial are examples of such distributions.

contributes to its future growth, since the base to which the percentage growth rate is applied is thereby increased by one unit. But the effect of a unit increase most likely diminishes as time goes on. In the simulation model in ch. 8, this pattern of diminishing effects is represented by an exponential decay. The simulation model showed that the greater the decay (or discount) rate, the greater the concavity of the distribution. This decay rate has a significant economic meaning: it measures the effect of growth momentum and of the stability of the economic environment. For example, when consumer brand loyalty is high and market share is stable, recent growth has a long carry-over effect. When consumer brand loyalty is low and market shares are unstable, recent growth has only a short-run effect upon the firm's future growth.

We are now able to derive analytically this effect of the decay rate, which we estimated by simulation in our earlier paper. In the next section, we summarize our analytical characterization of this model, leaving mathematical derivations to the Appendix.

In §3 we consider a second economic process that may contribute to the concavity of the distribution, namely, the effect of mergers and acquisitions. In ch. 10 we analyzed the effect of mergers and acquisitions on the concentration measure β. When this study was done, we had available only a very limited amount of data on mergers and acquisitions. We now have analyzed a more extensive set of data over a wider period, based on the Federal Trade Commission's report covering mergers and acquisitions of manufacturing and mining companies between 1948 and 1969. Contrary to our previous findings, the results of this analysis show that mergers and acquisitions have contributed to a noticeable extent to increasing the concentration measure and increasing the concavity of the firm-size distribution. We present our results of the merger analysis in §3.

2. Autocorrelated growth

The curvature of a firm-size distribution may be quantified in many different ways when we fit a curve to empirical data. It is possible to fit

$$\log s = \log M - \beta \log r + C(\log r)^2, \tag{3}$$

where the coefficient C depends on the curvature. The distribution is convex downward if C is positive and concave downward if C is negative.

Alternatively, we might select a set of pairs of ranks j_1 and j_2 and measure a representative value (mean, median, etc.) of the ratios for each pair:

$$R = s_l / (s_{j_1}^p s_{j_2}^{1-p}), \tag{4}$$

where s_l is the size of the rank l firm $(j_1 < l < j_2)$ and p $(0 < p < 1)$ satisfies $l = j_1^p j_2^{1-p}$. The denominator in (4) is the theoretical value for the size of the lth firm under the Pareto law, and the numerator is its actual value.

A stronger theoretical basis for quantifying and interpreting the curvature can be obtained by using a parameter that we introduced in ch. 8. In that chapter, we demonstrated by a simulation that the following process leads to a distribution which approximates the Pareto curve but with a curvature.

(1) At the start of the process, namely, at epoch 0 (a specific point in time), we have several firms in the population (an industry or an economy), each firm having a unit (sales) size[5] and weight 1. The weight is an indicator of the firm's growth potential relative to other firms. Placing several firms in the population is an initialization step for the model. It has been shown that after large numbers of epochs the resulting distribution is insensitive to the initialization (see ch. 8).

(2) At each epoch k, $k = 1, 2, 3, \ldots$, one unit of sales is given to a randomly selected firm.

(3) There is a constant probability α that the sales unit at epoch k goes to a new firm. If this happens, a new firm with size

[5]This can be stated in terms of a unit of assets, but we shall use sales units for the sake of ease in explanation.

1 is added to the population. This firm is treated as an old firm in all succeeding epochs.

(4) If the sales unit at epoch k does not go to a new firm, it goes to an old firm. The probability of a given old firm receiving the sales unit is proportional to its weight – that is, it is equal to the firm's weight divided by the sum of weights of all firms in the population. The size of the selected firm is increased by one unit.

(5) After a new firm is created or an old firm is selected among firms in the population, the weight of each old firm in the population is discounted by multiplying its current weight by γ. Then, if a new firm is created, its weight is set equal to 1; if an old firm is selected, its weight (after the above discounting) is increased by 1.

In ch. 8 we showed by simulation that the resulting firm-size distribution exhibits a downward concavity if $\gamma < 1$, is a straight line if $\gamma = 1$, and has a downward convexity if $\gamma > 1$. The curvature is greater as $|\gamma - 1|$ is greater. (For the case of $\gamma = 1$, namely, without discounting, see ch. 1). Therefore, γ may be used to quantify the curvature observed in an empirical distribution by fitting a theoretical distribution generated under a given γ.

The parameter γ has a simple economic interpretation. It indicates the relative significance of recent sales as compared with earlier sales in generating future sales. Both recent sales and prior sales contribute in generating future sales, but the latter contribute less than the former if $\gamma < 1$ and more than the former if $\gamma > 1$. Thus, if $\gamma < 1$, a firm experiencing growth recently has a better chance of growing than a firm of the same size that had its growth in earlier periods.

We now show these points analytically for the case where $0 < \gamma < 1$ (distribution concave downward).

Let W_k be the sum of weights of all firms in the population after step 5 for epoch k has been completed. Then,

$$W_{k+1} = \gamma W_k + 1, \tag{5}$$

as described in step 5. Then, for a sufficiently large k,

$$W_k = 1 + \gamma + \gamma^2 + \cdots + \gamma^k \cong \frac{1}{1-\gamma} \quad (0 < \gamma < 1).^6 \qquad (6)$$

Let us then consider a firm that was created at epoch k. The size of this firm at epoch k is 1 and so is its weight. Then, the probability that this firm will experience a sale at epoch $(k + 1)$ is the probability that the sale goes to an old firm $(1 - \alpha)$ times the probability that this particular firm experiences the sale, $[1 \div 1/(1 - \gamma) = 1 - \gamma]$, or a total probability, $(1 - \alpha)(1 - \gamma)$, which we denote by ξ. Thus, the probability that this firm does not get the sale at epoch $(k + 1)$ is $(1 - \xi)$.

Now assume that this firm did not get the sale at epoch $(k + 1)$. The weight is now discounted from 1 to γ for epoch $(k + 2)$. Thus, the probability of getting the sale at epoch $(k + 2)$ is $(1 - \alpha) \times (1 - \gamma)\gamma = \xi\gamma$, while the probability of not getting the sale at epoch $(k + 2)$ is $(1 - \xi\gamma)$. Similarly, the probability of not getting the sale at epoch $(k + 3)$ is $(1 - \xi\gamma^2)$.

Therefore, the probability that this firm will not get any sale after epoch k (that is, the probability u_1 that the size of this firm remains at 1 forever) is

$$u_1 = (1 - \xi)(1 - \xi\gamma)(1 - \xi\gamma^2) \ldots \qquad (7)$$

For γ close to but less than 1, $\xi = (1 - \alpha)(1 - \gamma)$ is close to 0, and so is $\xi\gamma^k$ for $k = 1, 2, \ldots.^7$ Then, we may use the approximation

$$\log(1 + x) \cong x, \qquad (8)$$

obtaining

$$\log u_1 \cong -\xi(1 + \gamma + \gamma^2 + \cdots) = -(1 - \alpha)(1 - \gamma)/(1 - \gamma), \qquad (9)$$

---

[6]More precisely, $W_k = 1 + \gamma + \gamma^2 + \cdots + m\gamma^k$, where m is the number of firms created at initialization with size 1 and weight 1. Clearly, m has only a small effect when k is large.

[7]If $\gamma(<1)$ is not close to 1, $\xi\gamma^k$ becomes close to zero except for the first few terms. In the simulation in ch. 8, it was noted that even a mild departure of γ from 1, such as 0.95, is sufficient to create a considerable curvature in the distribution.

or

$$u_1 \cong e^{-(1-\alpha)}. \tag{10}$$

In general, the probability of a firm with a present weight w not getting any sales in the future is easily shown to be

$$u_w \cong e^{-(1-\alpha)w} \tag{11}$$

by applying the above argument. Eq. (10) is a special case for $w = 1$.

The implication of the expression in (11) to the firm-size distribution is this: first, u_1 is the expected asymptotic proportion of the unit-sized firms among all firms. In (10), u_1 depends only on α and is independent of γ.

Next, consider a firm of size i, and weight w. Here,

$$w = \sum_{\tau=0}^{k} y(\tau)\gamma^\tau \tag{12}$$

where $y(\tau) = 1$ if the firm received the sale at epoch $(k - \tau)$ and $y(\tau) = 0$ otherwise. The sequence $y(0), y(1), \ldots$, should have exactly i 1's. Therefore, for any given history of a firm (given $\{y(\tau)\}$),

$$\frac{dw}{d\gamma} > 0. \tag{13}$$

But, the probability of this firm not receiving any more sales is given by (11), namely, $u = (1 - \xi w)(1 - \xi\gamma w)(1 - \xi\gamma^2 w) \cdots \cong e^{-(1-\alpha)w}$. Hence,

$$\frac{du}{dw} = \frac{du}{d\log u} \frac{d\log u}{dw} = u \sum_{\tau=0}^{\infty} -\xi\gamma^\tau / (1 - \xi\gamma^\tau w) < 0. \tag{14}$$

Thus, from (13) and (14),

$$\frac{du}{d\gamma} < 0. \tag{15}$$

This means that the probability of having further sales $(1 - u)$ for old firms becomes smaller as γ becomes smaller (namely, as discounting becomes heavier). Growth beyond size 1 becomes more and more difficult as weights are discounted more heavily (or as γ becomes smaller).

Furthermore, it can easily be seen from (12) that on the average as the size s of a firm increases, the effect of reducing γ on w becomes more severe since there are more terms in the expression for w that are reduced as γ becomes smaller. Thus, on the average, reducing γ has a greater effect on the probability of further growth as size s becomes larger. Therefore, not only is the probability of further growth for old firms decreased as γ becomes smaller, but also the decrease in probability is greater the larger the firm.

From this analysis, we can conclude that as γ becomes smaller the concavity of the firm-size distribution on a double-log scale is increased.

In the Appendix, we present a further analysis of the effect of γ on the concavity from the viewpoint of the mean and the variance of the distribution of firm sizes.

The empirical support for this autocorrelation between recent growth and future growth (in our analysis this means $\gamma < 1$) has been given in ch. 9. There, an excess growth ratio of a firm in 1954–58 over and above the average growth ratio of the economy is definitely carried over to the subsequent period in 1958–62 but only in proportion to the cube root of the original excess growth ratio.

Empirically, this decay effect appears to be quite plausible. When two firms of equal size are compared, the one that grew more recently seems to have a better chance for growth than the other whose growth occurred in the remote past. Growth momentum is not likely to change suddenly.

Thus, the existence of concavity in the firm-size distribution does not seem to be a good reason for abandoning the Pareto curve but a good reason for modifying the Pareto curve by introducing autocorrelation in the growth process.

3. Mergers and acquisitions

Let us now turn to the second economic factor that may have contributed to creating concavity in the firm-size distribution, namely, the effects of mergers and acquisitions.

In ch. 10, we analyzed the steady state distribution of firm sizes in an economy in the following two steps. A first process selects firms that disappear as a result of mergers and acquisitions. Then, the assets of all disappearing firms are pooled and allocated to surviving firms. This separation into two steps allowed us to analyze data that were not detailed enough to make a correlation analysis of the relation between the sizes of acquiring firms and the sizes of acquired firms.

The theoretical model to which the data were compared in ch. 10 was based on the following two assumptions: (1) the probability of a firm disappearing is independent of its size, and (2) the pooled assets of disappearing firms are allocated among surviving firms in such a way that the percentage increase due to allocation in the pooled size of the surviving firms in a given size range is independent of the size range. The empirical data we examined in ch. 10 were not extensive enough to detect any significant departure from the above assumptions. These data referred to mergers and acquisitions that occurred in 1956–57, when mergers and acquisitions were not as frequent as in the 1960s.

We obtained and examined more detailed data covering the period 1948–69.[8] After having analyzed this new set of data to retest the model, we found significant departures from the above

[8]There are a number of differences between the data in this chapter and those in ch. 10 that complicate our task of interpreting the findings. The present study covers a longer time period and more firms, enabling us, for example, to classify firms in 10 size categories instead of the four categories of the earlier study. Size is measured in the present study by assets and in the previous study by sales, but we do not believe that this difference changes matters very much. It is entirely possible that the Pareto curve provides a better approximation for the earlier period both because mergers were less frequent and because those that occurred were distributed in closer conformity to the assumptions of the model than the more frequent mergers of the 1960s were.

assumptions, sufficient to produce some curvature of the firm-size distribution.

Tables 11.1 and 11.2 are based on the Federal Trade Commission's publication *Large Mergers in Manufacturing and Mining, 1948–1969* (1970) and the *Fortune* data (1970a, 1970b) on assets of the largest industrial corporations as of December 31, 1969. In table 11.1, we assemble the basic data needed for estimating the "merger-free" distribution – the distribution which would be obtained from the actual 1969 data had no mergers and acquisitions occurred in the 1948–69 period. We can then see the effect of mergers and acquisitions on the firm-size distribution by comparing the (estimated) merger-free and the (actual) post-merger distributions. Table 11.1 also tests the assumption (1 above) that the probability of disappearance of firms by mergers and acquisitions is independent of the size.

In table 11.1, column 1 shows the frequencies of firms in the indicated asset size ranges as of December 31, 1969, based on the *Fortune* data. Columns 2, 3, and 4 show the frequencies of firms involved in mergers and acquisitions in 1948–69 as reported in the Federal Trade Commission report (1970). They are classified in three ways, based on postmerger size (col. 2), the premerger size of the acquiring firm (col. 3), and the premerger size of the acquired firm (col. 4).

Now suppose that all mergers in 1948–69 had occurred at once just before the close of the year 1969. This is not a totally unrealistic assumption, since by far the greatest portion of the mergers are concentrated in the last 2–3 years of the 20-year period. We can then calculate the size distribution of firms that would result if all such mergers and acquisitions were "undone" at a single moment in time, as if the merger movie were suddenly and swiftly run backward.

Take, for example, the size class 5 of firms with assets of $0.8–$1.6 billion. As of December 31, 1969, the *Fortune* data show 70 firms in this size class. We want to estimate what this number would be if mergers and acquisitions were undone all at once. To do so, we first note that in this size class there were 83 instances where firms acquired other firms and four where firms were

Table 11.1
Frequencies of firms classified by asset sizes.

Size class and asset size Range ($ billion)	Number of large industrial firms in Fortune 1 000 in 1969 (1)	Number of firms involved in mergers, 1948–69, classified by:			Estimated no. of firms if effects of mergers are eliminated (1) − (2) + (3) + (4) (5)	Rate of disappearance (4)/(5) (6)
		Postmerger size of combined firms (2)	Premerger size of acquiring firms (3)	Premerger size of acquired firms (4)		
(1) Over 12.8	2	–	–	–	2	0.00
(2) 6.4–12.8	7	4	4	–	7	0.00
(3) 3.2–6.4	11	14	11	–	8	0.00
(4) 1.6–3.2	33	36	31	1	29	0.03
(5) 0.8–1.6	70	94	83	4	63	0.06
(6) 0.4–0.8	95	147	127	9	84	0.11
(7) 0.2–0.4	137	184	172	20	145	0.14
(8) 0.1–0.2	214	206	171	83	262	0.32
(9) 0.05–0.1	262	159	160	139	402	0.35
	831	844	759	256	1 002	–
(10) Below 0.05	–	146	231	734	–	–
	–	990	990	990	–	–

Table 11.2
Aggregate assets of firms classified by asset sizes.

Size class and asset size range	Fortune 1 000 in 1969 (1)	Firms involved in mergers, 1948–69			Estimated without mergers (1) − (2) + (3) + (4) (5)	Surviving firms (5) − (4) (6)	Acquired firm's assets classified by size of acquiring firms (7)	Rate of growth (7)/(6) (8)
		Postmerger firms (2)	Acquiring firms (3)	Acquired firms (4)				
(1) Over 12.8	32.36	–	–	–	32.36	32.36	–	0.00
(2) 6.4–12.8	54.61	39.09	38.89	–	54.41	54.41	0.20	0.00
(3) 3.2–6.4	49.86	56.87	45.20	–	38.19	38.19	0.91	0.02
(4) 1.6–3.2	72.56	78.03	69.32	1.85	65.70	63.85	4.95	0.08
(5) 0.8–1.6	82.07	106.66	93.65	4.23	73.29	69.06	8.52	0.12
(6) 0.4–0.8	53.23	83.78	71.97	5.24	46.66	41.42	10.28	0.25
(7) 0.2–0.4	38.71	54.67	50.74	6.11	40.89	34.78	11.43	0.33
(8) 0.1–0.2	29.69	30.76	24.95	11.39	35.27	23.88	7.77	0.33
(9) 0.05–0.1	18.54	12.21	11.80	9.53	27.66	18.13	5.12	0.28
	431.63	462.07	406.52	38.35	444.43	376.08	49.18	–
(10) Below 0.05	–	5.51	6.18	16.53	–	–	5.70	–
	–	467.58	412.70	54.88	–	–	54.88	–

acquired by other firms. On the other hand, there were 94 merger and acquisition instances where firms, after merger, belonged to this size class. Hence, there was a net gain of seven $(94 - 83 - 4 = 7)$ in the number of firms in this class by reason of mergers and acquisitions. Since the actual number of firms in this class in 1969 is 70, we estimate that $70 - 7 = 63$ would have been in existence in 1969 had there been no mergers, as stated in column 5. By taking the cumulative frequencies based on the data in column 5 and plotting the result in fig. 1 as a broken curve, we obtained an estimated rank-size relationship when all mergers and acquisitions are "undone." We shall come back to this point again shortly.

Finally, the last column, column 6, shows the percentage rate of disappearance of firms in each size class by mergers and acquisitions over the 20-year period. From this we can conclude that smaller firms have a higher chance of being absorbed by mergers and acquisitions, contrary to the assumption (1) on p. 205 of equal probability of disappearance.

To test the assumption (2), on the size independence of growth rate by mergers, we prepared table 11.2. Column 1 was taken from the *Fortune* data, adding the assets of all firms belonging to a given size class. Columns 2, 3, and 4 were prepared in the same way as in table 11.1 except that they refer to aggregate assets instead of frequencies.

Consider, again, the size class 5. The table shows that the aggregate assets of the postmerger firms belonging to this class are $106.66 billion while the aggregate assets of the acquiring and acquired companies are $93.65 and $4.23 billion, respectively. Thus, as a result of mergers and acquisitions, this size class gained $8.78 billion in aggregate assets $(= 106.66 - 93.65 - 4.23)$. Since the actual aggregate assets based on the *Fortune* data are $82.07 billion, we estimate the "mergerless" aggregate size of this class to be $73.29 billion, as shown in column 5. Other figures in this column were prepared in the same way.

From data in column 5, we eliminated the aggregate assets of acquired firms belonging to the class (col. 4). The result is shown in column 5. We would like to see how these aggregate assets of acquired firms are redistributed among the surviving firms. Column 7 shows the distribution of the aggregate assets of acquired firms based on the premerger size of the *acquiring* firms by which the acquired firms were absorbed. Thus, for example, the size class 5 increased its aggregate assets by 12 percent as a result of reallocation of assets of acquired firms. From the data in the last column, it seems clear that the assumption (2) of size-independent growth rate by reallocation is not satisfied.[9]

What then is the effect of mergers and acquisitions on the firm-size distribution? To show this effect graphically, we prepared from column 5 of table 11.1 the broken curve in fig. 11.1. The broken curve indicates an estimated firm-size distribution when the effects of mergers and acquisition are eliminated. This was superimposed on the actual postmerger size distribution given by the solid curve in fig. 11.1.

The broken line was obtained by connecting the sizes of the tenth and eight hundred thirty-first firms in the "mergerless" distribution in order to make it comparable to the solid line indicating the theoretical distribution for postmerger data.

Comparing the two distributions, shown by the solid and broken lines, respectively, we note the following two significant points. First, the slope of the solid line is steeper than the slope of the broken line. As discussed in ch. 10, this slope measures the degree of concentration.[10] Therefore, we can conclude that mergers and acquisitions in the 1948–69 period increased the degree of concentration.[11]

[9]To some extent, the conclusions from tables 11.1 and 11.2 are affected by the assumption that all mergers and acquisitions have taken place at once. A year-by-year analysis would indicate somewhat smaller differences among the rates of disappearance across the classes and among the rates of growth by asset reallocation across the classes.

[10]It can be seen from eq. (1) that if the rank, r, is increased to jr, the size, s, is reduced to s/j^β, indicating that the larger the β, the greater the relative size of a larger firm over a small firm when the ranks of the two firms are fixed. It is clear from eq. (2) that this β is the slope of the double-log distribution.

[11]For the "mergerless" data, the size of the tenth firm is estimated by interpolation to be \$5.706 billion and the size of the eight hundred thirty-first firm is estimated to be \$0.064 billion from the data in column 5 of table 11.1.

The second important implication of table 11.1 is that the concavity of the distribution is increased as a result of mergers and acquisitions. The vertical deviation between the curve (actual or estimated) and the line (theoretical) is much wider for the postmerger distribution than for the "mergerless" distribution.[12]

4. Summary and conclusions

We have presented two different economic explanations for the concavity of the double-log firm-size distributions as observed in empirical data, one based on the autocorrelation of growth and the other based on mergers and acquisitions.

Since both explanations are not only economically plausible but also supported by empirical data, it is likely that both factors have contributed to the concavity of the firm-size distribution. Whether there are other contributing factors or whether these two factors are sufficient to eliminate most of the concavity is difficult to determine. The degrees of autocorrelation reflected in γ would have to be determined independently from the observed curvature of the empirical distribution before we could answer this question. At present, we have no such independent estimate of γ.

However, the analysis in this chapter does indicate the advantage of adjusting the Pareto curve for these additional economic factors instead of abandoning it by reason of the concavity in the observed distribution.

From the standpoint of statistical curve fitting to empirical distributions, we may find other distributions to be superior to

---

Connecting these points, β is estimated at 1.015. For postmerger data, the size of the tenth firm is $6.146 billion and the size of the eight hundred thirty-first firm is $0.050 billion from the *Fortune* data. Connecting these points, β is estimated to be 1.089, an increase of 7 percent in the slope as a result of mergers and acquisitions.

[12]The greatest size variance (the ratio of actual over theoretical size) is approximately 1.6 for the "mergerless" data and 2.1 for the postmerger data.

the modified Pareto curve. However, it is not our sole objective to find a distribution that can best fit empirical data. What we are after is a model that has sound economic support and justification and that can still give a reasonably good fit to the data. The fact that the Pareto curve frequently approximates not only firm-size data but also a variety of economic and noneconomic skewed distributions is an important advantage in retaining it as a first approximation, instead of abandoning it.

5. Appendix

Consider a firm which was created at epoch k as a result of getting one unit of sale. The probability of getting a sale at epoch $(k + 1)$ is ξ. Its weight is discounted from 1 to γ at epoch $(k + 2)$. Hence, if the firm fails to receive the sale at epoch $(k + 1)$, the probability of a sale at epoch $(k + 2)$ is $\xi\gamma$. On the other hand, if the firm experiences a sale at epoch $(k + 1)$, its weight is changed to $(\gamma + 1)$; hence the probability of a sale at epoch $(k + 2)$ is $(\xi\gamma + \xi)$. Note that in either case the original sale contributes $\xi\gamma$ to the probability of a sale at epoch $(k + 2)$ regardless of whether there was a sale at epoch $(k + 1)$.

In general, a sale at epoch k contributes to the probability of a sale at epoch $(k + \tau)$ by the amount $\xi\gamma^{\tau-1}$. Thus, if we focus only on the contribution by the *original* sale, this contribution decreases geometrically from ξ for epoch $(k + 1)$, $\xi\gamma$ for epoch $(k + 2)$, $\xi\gamma^2$ for epoch $(k + 3)$, and so on. Note that the events at each epoch (based only on the contribution to probability by the original sale) are mutually independent. Therefore, we have a sequence of independent Bernoulli trials where the probability of success decreases geometrically. The sum of sales generated $N_1 = X_1 + X_2 + \cdots$, where X is 1 if a sale is generated at epoch $(k + \tau)$ and 0 otherwise, has the generating function

$$P_1(x) = [1 - \xi(1 - x)][1 - \xi\gamma(1 - x)][1 - \xi\gamma^2(1 - x)] \ldots .$$
(16)

Note that $P_1(1) = 1$. The subscript 1 in N_1 and P_1 is used to indicate that this expression involves only the sales generated from the original sales.

Hence, the derivative of $P_1(x)$ with respect to x is

$$P_1'(x) = \frac{P_1(x)}{1 - \xi(1 - x)} + \frac{P_1(x)}{1 - \xi\gamma(1 - x)} + \frac{P_1(x)\xi\gamma^2}{1 - \xi\gamma^2(1 - x)} \cdots$$

$$= P_1(x) \sum_{\tau=0}^{\infty} \frac{\xi\gamma^{\tau}}{1 - \xi\gamma^{\tau}(1 - x)}. \tag{17}$$

The expected value of N_1, $E(N_1)$, is then given by

$$E(N_1) = P_1'(1) = P_1(1)\xi \sum_{\tau=0}^{\infty} \gamma^{\tau} = (1 - \alpha)(1 - \gamma)$$

$$\times \frac{1}{1 - \gamma} = 1 - \alpha. \tag{18}$$

The second derivative, $P_1''(x)$, is from (17):

$$P_1''(x) = P_1'(x) \sum_{\tau=0}^{\infty} \frac{\xi\gamma^{\tau}}{1 - \xi\gamma^{\tau}(1 - x)}$$

$$- P_1(x) \sum_{\tau=0}^{\infty} \frac{\xi^2\gamma^{2\tau}}{[1 - \xi\gamma^{\tau}(1 - x)]^2}. \tag{19}$$

But from, (17)

$$\sum_{\tau=0}^{\infty} \frac{\xi\gamma^{\tau}}{1 - \xi\gamma^{\tau}(1 - x)} = \frac{P_1'(x)}{P_1(x)}. \tag{20}$$

Substituting this in (19),

$$P_1''(x) = \frac{[P_1'(x)]^2}{P_1(x)} - P_1(x) \sum_{\tau=0}^{\infty} \frac{\xi^2\gamma^{2\tau}}{[1 - \xi\gamma^{\tau}(1 - x)]^2}. \tag{21}$$

Using the result in (18),

$$P_1''(1) = (1 - \alpha)^2 - \xi^2 \sum_{\tau=0}^{\infty} \gamma^{2\tau}$$

$$= (1 - \alpha)^2 - (1 - \alpha)^2(1 - \gamma)^2 \frac{1}{1 - \gamma^2}$$

$$= (1 - \alpha)^2 \frac{2\gamma}{1 + \gamma}. \tag{22}$$

Now the variance of N_1, var (N_1), is given (see Feller, (1968, p. 266)) by

$$\text{var}(N_1) = P''_1(1) + P'_1(1) - [P'_1(1)]^2$$

$$= (1 - \alpha)^2 \frac{2\gamma}{2 + \gamma} + (1 - \alpha) - (1 - \alpha)^2$$

$$= (1 - \alpha)\left[1 - \frac{(1 - \alpha)(1 - \gamma)}{1 + \gamma}\right]. \tag{23}$$

Thus, var $(N_1) \to 1 - \alpha$ as $\gamma \to 1$ and var $(N_1) \to \alpha(1 - \alpha)$ as $\gamma \to 0$.

Note that if γ or α is close to 1, $\xi = (1 - \alpha)(1 - \gamma)$ is close to 0; therefore, applying the approximation (8) to (16), we obtain.

$$\log P_1(x) = -(1 - x)\xi(1 + \gamma + \gamma^2 \ldots) = -(1 - x)(1 - \alpha) \tag{24}$$

or

$$P_1(x) = e^{-(1-\alpha)(1-x)}, \tag{25}$$

which is a generating function of a Poisson distribution with mean $1 - \alpha$. However, regardless of the value of γ or α, the mean and the variance of N_1 given in (18) and (23) are exact and not approximations.

Now let us take into account the fact that the sales generated by the original sale also contribute to generating further sales. Thus, we have a branching process, each new sale having the potential of generating further sales. We may call the original sale a zero generation sale and a sale generated as a result of an nth generation sale an $(n + 1)$ generation sale.

Strictly speaking, the independence assumption needed for the application of theorems in branching processes is not satisfied here. For example, at epoch $(k - \tau)$ the probability of the original sale generating another one is $\xi\gamma^{\tau-1}$. At the same time, if the original sale has generated another sale at epoch $(k + 1)$, the latter also has a probability of generating another sale at epoch $(k + \tau)$ equal to $\xi\gamma^{-2}$. The two events are not independent. It is not possible for the original sale and the first-generation sale both to generate sales, because only one unit of sales occurs at epoch $(k + \tau)$. However, by reducing the time interval between epochs and making a corresponding adjustment to γ, ξ can be made as small as we want. Hence, the probability of having two or more sales occurring in a given epoch may be considered to be negligible.

Let N_j be the size of the jth-generation sales and P_j be the generating function of its probability distribution. Since there is only one original sale, we set $N_0 = 1$. The generating function of N_1 has already been analyzed above.

The jth-generation sales, N_j, may be divided into N_1 groups based on their ancestor in the first generation. The number in each group has the same probability distribution as N_{j-1}. Since they may be considered mutually independent approximately, the generating function P_j is given by a compound function (see Feller (1968, pp. 295–96)),

$$P_j(x) = P(P_{j-1}(x)). \tag{26}$$

From this, we obtain

$$P'_j(x) = P'_1(x)P'_{j-1}(x) = (P'_1(x))^j, \tag{27}$$

$$P''_j(x) = j(P'_1(x))^{j-1}P''_1(x). \tag{28}$$

Hence, $E(N_j)$ and var (N_j) are

$$E(N_j) = P'_j(1) = (1 - \alpha)^j, \tag{29}$$

$$\begin{aligned}
\text{var}\,(N_j) &= P''_j(1) + P'_j(1) - (P'_j(1))^2 \\
&= j(P'_1(1))^{j-1}P''_1(1) + (1-\alpha)^j - (1-\alpha)^{2j} \\
&= j(1-\alpha)^{j-1}\left[(1-\alpha)^2 - (1-\alpha)^2\frac{1-\gamma}{1+\gamma}\right] \\
&\quad + (1-\alpha)^j - (1-\alpha)^{2j} \\
&= j(1-\alpha)^{j+1}\frac{2\gamma}{1+\gamma} + (1-\alpha)^j - (1-\alpha)^{2j}. \tag{30}
\end{aligned}$$

Now let V_j be

$$V_j = N_0 + N_1 + \cdots + N_j, \tag{31}$$

and let

$$V = \lim_{j \to \infty} V_j, \tag{32}$$

namely, the ultimate size of a firm. The generating function of V must satisfy

$$Q(x) = x/P(Q(x)) \tag{33}$$

(Feller (1968, p. 298)). Hence,

$$Q'(x) = P(Q(x)) + sP'(Q(x))Q'(x), \tag{34}$$

$$Q'(x) = P(Q(x)/[1 - xP'(Q(x))], \tag{35}$$

$$E(V) = Q'(1) = \frac{1}{1 - (1 - \alpha)} = \frac{1}{\alpha}. \tag{36}$$

Also from (34),

$$Q''(x) = 2P'(Q(x))Q'(x) + sP''(Q(x))(Q'(x))^2$$
$$+ sP'(Q(x))Q''(x), \tag{37}$$

$$Q''(x) = \frac{2P'(Q(x))Q'(x) + xP''(Q(x))(Q'(x))^2}{1 - xP'(Q(x))}. \tag{38}$$

Hence,

$$Q''(1) = \frac{1}{\alpha} \left[\frac{2(1 - \alpha)}{\alpha} + \frac{(1 - \alpha)^2}{\alpha^2} \frac{2\gamma}{1 + \gamma} \right]. \tag{39}$$

The variance of V is

$$\begin{aligned}
\text{var}(V) &= Q''(1) + Q'(1) - [Q'(1)]^2 \\
&= \frac{1}{\alpha} \left[\frac{2(1 - \alpha)}{\alpha} + \frac{(1 - \alpha)^2}{\alpha^2} \frac{2\gamma}{1 + \gamma} + 1 - \frac{1}{\alpha} \right] \\
&= \frac{1 - \alpha}{\alpha^2} + \frac{(1 - \alpha)^2}{\alpha^3} \frac{2\gamma}{1 + \gamma}. \tag{40}
\end{aligned}$$

Thus, for $0 < \alpha < 1$, the growth process of a firm dies out with probability 1 since $E(v) = 1/\alpha < \infty$. Furthermore, we note that $E(V)$ depends only on α and is independent of γ. In addition, var (V) is decreased as γ is decreased (namely, heavier discounting). However, since the total number of firms generated depends only on α and epoch k, the tail end of the firm-size distribution for the largest rank is fixed for given α and k and is independent of γ. Thus, in order to reduce the variance of V, the sizes of the above-average firms must be reduced and the sizes of below-average firms must be increased, thus creating more curvature in consistency with our earlier observations.

References

Adelman, I. G., "A Stochastic Analysis of the Size-Distribution of Firms," *Journal of the American Statistical Association*, 53, 1958, p. 893.

Aitchison, J. and J. A. C. Brown, *The Lognormal Distribution*. Cambridge: Cambridge University Press, 1957.

Artin, Emil, *The Gamma Function*. Translated by Michael Butler. New York: Holt, Rinehart & Winston, 1964.

Bain, J. S., *Barriers to New Competition*. Cambridge: Harvard University Press, 1956.

Banet, L., "Evolution of the Balmer Series," *American Journal of Physics*, 34, 1966, pp. 496–503.

Bower, G. H. and T. R. Trabasso, "Concept Identification," in R. C. Atkinson (ed.), *Studies in Mathematical Psychology*. Stanford: Stanford University Press, 1964, pp. 32–94.

Champernowne, D. G., "A Model of Income Distribution," *Economic Journal*, 63, June 1953, pp. 318–51.

Cohen, Joel E., *A Model of Simple Competition*. Cambridge: Harvard University Press, 1966.

Collins, N. R. and L. E. Preston, "The Size Structure of the Largest Industrial Firms 1909–58," *American Economic Review*, 51, 1961, pp. 986–1011.

Crum, W. L., *Corporate Size and Earning Power*. Cambridge: Harvard University Press, 1939.

Darwin, J. H., "Population Differences between Species Growing According to Simple Birth and Death Processes," *Biometrika*, 40, 1953, pp. 370–382.

Davis, Harold T., *The Analysis of Economic Time Series*. Principia Press, 1941.

Engwall, Lars, *Size Distribution of Firms*. Department of Business Administration, Stockholm University, 1970.

Estes, W. K., "Growth and Functions of Mathematical Models for Learning," in W. Dennis et al., *Current Trends in Psychological Theory*. Pittsburgh: University of Pittsburgh Press, 1961, pp. 134–151.

Federal Trade Commission, *Report on Corporate Mergers and Acquisitions*. Washington: Government Printing Office, May 1955.

——, *Large Mergers in Manufacturing and Mining, 1948–1969*. Statistical Report No. 5. Washington: Government Printing Office, February 1970.

Feldman, J., "Simulation of Behavior in the Binary Choice Experiment," in E. A. Feigenbaum and J. Feldman (eds.), *Computers and Thought*, New York: McGraw–Hill, 1964, pp. 329–346.

Feller, W., *An Introduction to Probability Theory and Its Applications*. Volume 1, 1st edition. New York: John Wiley & Sons, 1950.

——, *An introduction to Probability Theory and Its Applications*, Volume 1, 3rd edition. New York: John Wiley & Sons, 1968.

Fortune, "The Fortune Directory of the 500 Largest U.S. Industrial Corporations," *Fortune*, 54, July 1956, Supplement.

———, "The Fortune Directory of the 500 Largest U.S. Industrial Corporations," *Fortune*, 63, July 1961, Supplement.

———, "The Fortune Directory of the 500 Largest U.S. Industrial Corporations," *Fortune*, 81, May 1970a.

———, "The Fortune Directory of the Second 500 Largest U.S. Industrial Corporations," *Fortune*, 81, June 1970b.

Friedman, Milton, *Essays in Positive Economics*. Chicago: University of Chicago Press, 1953.

Good, I. J., "The Population Frequencies of Species and the Estimation of Population Parameters," *Biometrika*, 40, 1953, pp. 237–264.

Goodman, N., "The Test of Simplicity," *Science*, 176, 1958, pp. 1064–1069.

Gregg, L. W. and H. A. Simon, "Process Models and Stochastic Theories of Simple Concept Formation," *Journal of Mathematical Psychology*, 4, 1967a, pp. 246–276.

——— and H. A. Simon, "An Information Processing Explanation of One-Trial and Incremental Learning," *Journal of Verbal Learning and Verbal Behavior*, 6, 1967b, pp. 780–787.

———, A. P. Chenzoff and K. Laughery, "The Effect of Rate of Presentation, Substitution and Mode of Response in Paired-Associate Learning," *American Journal of Psychology*, 76, 1963, pp. 110–115.

Hanley, Miles L., *Word Index to James Joyce's Ulysses*. Madison: University of Wisconsin Press, 1937.

Hanson, R. N., *Patterns of Discovery*. Cambridge: Cambridge University Press, 1961.

Haran, E. G. P. and Daniel R. Vining, Jr., "On the Implications of a Stationary Urban Population for the Size Distribution of Cities," *Geographical Analysis*, 5, 1973, pp. 296–308.

Hart, P. E. and E. H. Phelps Brown, "The Sizes of Trade Unions: A Study in the Laws of Aggregation," *Economic Journal*, 67, March 1957, pp. 1–15.

———, and S. J. Prais, "The Analysis of Business Concentration," *Journal of Royal Statistical Society*, Part 2, 119, 1956, pp. 150–91.

Hill, Bruce M., "The Rank-Frequency Form of Zipf's Law," *Journal of American Statistical Association*, 69, 1974, pp. 1017–1026.

Ijiri, Yuji and Herbert A. Simon, "Business Firm Growth and Size," *American Economic Review*, 54, March 1964, pp. 77–89. (Chapter 8).

——— and ———, "A Model of Business Firm Growth," *Econometrica*, 35, April 1967, pp. 348–355. (Chapter 9).

——— and ———, "Effects of Mergers and Acquisitions on Business Firm Concentration," *Journal of Political Economy*, 79, March/April 1971, pp. 314–322. (Chapter 10).

——— and ———, "Interpretations of Departures from the Pareto Curve Firm-Size Distributions," *Journal of Political Economy*, 82, March/April 1974, pp. 315–331. (Chapter 11).

——— and ———, "Some Distributions Associated with Bose–Einstein Statistics," *Proceedings of National Academy of Sciences*, 72, May 1975a, pp. 1654–1657. (Chapter 4).

_____ and _____, "Properties of the Yule Distribution," Working Paper, Carnegie-Mellon University, 1975b. (Chapter 3).

Jeffreys, H. and D. Wrinch, "On Certain Fundamental Principles of Scientific Inquiry," *Philosophical Magazine*, 42, 1921, pp. 369–390.

Kendall, David G., "On Some Modes of Population Growth Leading to R. A. Fisher's Logarithmic Series Distribution," *Biometrika*, 35, 1948, p. 6.

Leavens, Dickson H., "Communication," *Econometrica*, 21, 1953, p. 630.

Levine, M., "Hypothesis Behavior by Humans During Discrimination Learning," *Journal of Experimental Psychology*, 71, 1966, pp. 331–338.

Lydall, H. F., "The Growth of Manufacturing Firms," *Bulletin of the Oxford University Institute of Statistics*, 21, 1959, pp. 85–111.

Mandelbrot, B., "An Informational Theory of the Statistical Structure of Language," in Willis Jackson (ed.), *Communication Theory*. London: Butterworths, 1953, pp. 486–502.

_____, "On Recurrent Noise-Limiting Coding," in *Proceedings of Symposium on Information Networks*. New York: Polytechnic Institute of Brooklyn, 1954, pp. 205–222.

_____, "A Note on a Class of Skew Distribution Functions," *Information and Control*, 2, 1959, pp. 90–99.

Mansfield, Edwin, "Entry, Gibrat's Law, Innovation, and the Growth of Firms," *American Economic Review*, 52, December 1962, pp. 1023–1051.

Miller, G. A., E. B. Newman and E. A. Friedman, "Length-Frequency Statistics for Written English," *Information and Control*, 1, 1958, pp. 370–389.

Moody, *Moody's Industrial Manual*. New York: Moody's Investors Service, 1956, 1957.

Moore, Fred, "Economies of Scale," *Quarterly Journal of Economics*, 73, 1959, pp. 232–245.

Penrose, Edith T., *The Theory of the Growth of the Firm*. New York: Wiley, 1959.

Popper, K. R., *The Logic of Scientific Discovery*. New York: Basic Books, 1961.

Postman, L., "One-Trial Learning," in C. F. Cofer and B. S. Musgrave (eds.), *Verbal Behavior and Learning*. New York: McGraw–Hill, 1963, pp. 295–321.

Rapoport, A., "Comment: The Stochastic and the 'Teleological' Rationales of Certain Distributions and the So-Called Principle of Least Effort," *Behavioral Science*, 2, 1957, pp. 147–161.

Robinson, Joan, *The Economics of Imperfect Competition*. London: Macmillan, 1942.

Rock, I., "The Role of Repetition in Associative Learning," *American Journal of Psychology*, 70, 1957, pp. 186–193.

Rowthorn, Robert, in Collaboration with Stephen Hymer, *International Big Business, 1957–67*. Cambridge: Cambridge University Press, 1971.

Savage, Leonard J., *The Foundations of Statistics*. New York: John Wiley & Sons, 1954.

Simon, Herbert A., "Prediction and Hindsight as Confirmatory Evidence," *Philosophy of Science*, 22, 1955a, pp. 227–230.

_____ "On a Class of Skew Distribution Functions," *Biometrika*, 52, December 1955b, pp. 425–440. (Chapter 1).

———, "Amounts of Fixation and Discovery in Maze Learning Behavior," *Psychometrika*, 22, 1957, pp. 261–268.

———, "Some Further Notes on a Class of Skew Distribution Functions," *Information and Control*, 3, 1960, pp. 80–88. (Chapter 2).

———, "A Note on Mathematical Models for Learning," *Psychometrika*, 27, 1962, pp. 417–418.

———, "Scientific Discovery and the Psychology of Problem Solving," in R. Colodny (ed.), *Mind and Cosmos*. Pittsburgh: University of Pittsburgh Press, 1966, pp. 22–40.

———, "On Judging the Plausibility of Theories," in B. van Rootselaar and J. F. Staal (eds.), *Logic, Methodology and Philosophy of Sciences*, Vol. III. Amsterdam: North-Holland, 1968. (Chapter 6).

——— and Charles P. Bonini, "The Size Distribution of Business Firms," *American Economic Review*, 48, September 1958, pp. 607–617. (Chapter 7).

——— and T. A. Van Wormer, "Some Monte Carlo Estimates of the Yule Distribution," *Behavioral Science*, 8, July 1963, pp. 203–210. (Chapter 5).

Singh, A. and G. Whittington, *Growth, Profitability and Valuation*. Cambridge: Cambridge University Press, 1968.

Steindl, J., *Random Process and the Growth of Firms: A Study of the Pareto Law*. New York: Hafner, 1965.

Thorndike, Edward L., "On the Number of Words of Any Given Frequency of Use," *Psychological Record*, 1, 1937, p. 399.

Titchmarsh, E. C., *The Theory of Functions*, 2nd edition. Oxford: Clarendon Press, 1939.

Underwood, B. J. and G. Keppel, "One-Trial Learning," *Journal of Verbal Learning and Verbal Behavior*, 3, 1964, pp. 385–396.

U.S. Congress, House, Select Committee on Small Business, *Mergers and Super-concentration: Acquisitions of 500 Largest Industrial and 50 Largest Merchandising Firms*. 87th Congress. Washington: Government Printing Office, November 8, 1962.

U.S. Congress, Senate, Committee on the Judiciary, Subcommittee on Antitrust and Monopoly, *Economic Concentration*. Hearings, July and September 1964, Part 1. Washington: Government Printing Office, 1964.

Viner, J., "Cost Curves and Supply Curves," in *Zeitschrift f. Nationalökon*. 1931. Reprinted in G. J. Stigler and K. E. Boulding (eds.), *AEA Readings in Price Theory*. Homewood, IL: R. D. Irwin, 1952, pp. 198–232.

Vining, Daniel R., Jr., *Models of Urban and Spatial Concentration*. Ph.D. Dissertation, School of Urban and Public Affairs, Carnegie-Mellon University, 1975.

Walters, A. A., "Production and Cost Functions: an Econometric Survey," *Econometrica*, 31, January–April, 1963, pp. 1–66.

Wedervang, Froystein, *Development of a Population of Industrial Firms*. Oslo: Universitetsforlaget, 1965.

Wold, H. O. A. and P. Whittle, "A Model Explaining the Pareto Law of Wealth Distribution," *Econometrica*, 25, October 1957, pp. 591–595.

Yule, G. Udny, "A Mathematical Theory of Evolution, Based on the Conclusions of Dr. J. C. Willis, F.R.S.," *Philosophical Transactions*, B. 213, 1924, pp. 21–83.

———, *The Statistical Study of Literary Vocabulary*. Cambridge: Cambridge University Press, 1944.

Zipf, G. K., *Human Behavior and the Principle of Least Effort*. Reading, MA: Addison Wesley, 1949.

Author index

Subject index